Management of Deteriorating Concrete Structures

Also available from Taylor & Francis

Reynolds's Reinforced Concrete Designer's Handbook 11th edition

C. Reynolds et al.

Hb: ISBN 978–0–419–25820–9
Pb: ISBN 978–0–419–25830–8

Concrete Structures 3rd edition

A. Ghali et al.

Hb: ISBN 978–0–415–24721–4

Corrosion of Steel in Concrete 2nd edition

J. Broomfield

Hb: ISBN 978–0–415–33404–4

Protection and Repair of Concrete Structures

H. Davies

Hb: ISBN 978–0–415–26725–0

Testing of Concrete Structures 4th edition

J. Bungey et al.

Hb: ISBN 978–0–415–26301–6

Information and ordering details

For price availability and ordering visit our website
www.tandfbuiltenvironment.com/
Alternatively our books are available from all good bookshops.

Management of Deteriorating Concrete Structures

George Somerville

Routledge
Taylor & Francis Group

LONDON AND NEW YORK

First published 2008 by Taylor & Francis

2 Park Square, Milton Park, Abingdon, Oxfordshire OX14 4RN
52 Vanderbilt Avenue, New York, NY 10017

Routledge is an imprint of the Taylor & Francis Group, an informa business

First issued in paperback 2019

Copyright © 2008 George Somerville

Typeset in Sabon by
Integra Software Services Pvt. Ltd, Pondicherry, India

British Library Cataloguing in Publication Data
A catalogue record for this book is available from the British Library

Library of Congress Cataloging in Publication Data
Somerville, G.
    Management of deteriorating concrete structures / George Somerville.
        p. cm.
    Includes bibliographical references and index.
    ISBN 978-0-415-43545-1 (hardback : alk. paper)
    1. Concrete--Deterioration. 2. Concrete
    construction--Maintenance and repair. I. Title.
    TA440.S637 2007
    624.1'8340288--dc22
    2007011262

ISBN13: 978-0-415-43545-1 (hbk)
ISBN13: 978-0-367-38738-9 (pbk)

Contents

Figures

Tables

Preface

The motivation for writing this book came mainly from the author's involvement in three successive EC-funded projects over a 10-year period, all relating to durability, assessment, repair and management of deteriorating concrete structures. The acronyms for these are BRITE 4062, CONTECVET and REHABCON, and all three are well referenced in this book. BRITE 4062 was concerned with the science of deteriorology, CONTECVET with the assessment of the effects that deterioration might produce and REHAB-CON with how this new information might fit into existing asset management systems. The partnership, from Spain, Sweden and the UK, was a mix of scientists and engineers, of researchers and contractors, with the owner contingent increasing as the work progressed. Indeed, an unforeseen benefit from this 10-year collaboration was a proper understanding, by the technocrats, of the real needs of owners on a day-to-day basis and of how best to meet these.

It was here that the concept of progressive screening, used in this book, first emerged, while keeping the approach as simple and practical as possible, commensurate with the quantity and quality of the information likely to be available for most assessments. The approach was essentially deterministic, in modifying design equations to allow for deterioration, while introducing the results from recent relevant research. Owners valued this approach, since it gave a direct comparison of how the present and future states compared with that provided for in the original design.

Over the years, different asset management systems have emerged for different types of structure in different countries – driven by need and often in splendid isolation with little technology transfer. The book gives some examples of these. This evolution continues apace, resulting in more rigorous computer-based management systems, with input obtained from more sophisticated measurement techniques. In parallel with this, research on deteriorology has moved forward from understanding the basic mechanisms involved towards an appreciation of their effects on the performance of structures. Here, the general trend has been to go back to first principles based on reliability theory and probabilistic methods. While this is understandable and commendable, the practical lack of sufficient data remains a problem, as does a real appreciation of deterioration on all the key action effects which can only come from physical testing. In short, research outputs are not yet in the ideal form to meet the real needs of owners in their management role, i.e. in deciding on whether remedial action is necessary, and, if so, what is the best option. Repair and remedial action is a whole world in itself, with so many options now available that selection is something of a problem. Thankfully this is now changing, with the evolution of a more scientific approach to evaluating repair options, and, in this book, most weight is put on the approach typified by EN1504, while drawing attention to feedback on the performance of remedial measures in the field.

Plainly, there is a need to fill in the gaps identified above and to integrate the whole asset management process at an effective and practical level. Without pretending to have all the answers, I would like to think that this book might help a little, while paving the way for future developments. The scope of the subject is so large and diverse that the emphasis has been put on principles and procedures, while making reference to sound guidance already on record, for the detail in individual sectors.

George Somerville
Old Windsor, Berks
May 2007

Acknowledgements

The impetus for writing this book came from the author's involvement in three consecutive EC-funded projects. These involved partners from Spain, Sweden and the UK. Significant technical input came from Geocisa and IETcc in Spain; from CBI, Skanska, Cementa AB and Lund Technical University in Sweden and from the British Cement Association, BRE and TRL in the UK. Valuable input also came from major owners in all the three countries, covering a range of structures including bridges, buildings, car parks, and a variety of civil engineering installations. The individuals involved are too numerous to mention but they know who they are. The author is extremely grateful to all these people and organisations for the positive and pleasant collaboration over a 10-year period, which not only provided much useful information of immediate value to the owner-partners but also led to the concept of progressive screening which is central to this book.

Many organisations have produced guidance on different facets of asset management. For much of the detail, the author has drawn freely on this, particularly outputs from the Concrete Society, *fib*, ACI, BRE, BSI, the Concrete Repair Association, and the Concrete Bridge Development Group. These references are all fully listed, with acknowledgements where figures and tables have been reproduced. The existence of these detailed documents has permitted the author to focus on the principles and procedures involved.

It is perhaps invidious to mention individuals, running the risk of the sin of omission. However, some are particularly deserving. Former colleagues, over a 40-year period, at the British Cement Association and the Cement and Concrete Association come into this category. In particular, Dr Mike Webster, who worked closely with the author on the EC-funded projects and went on to widen the scope by completing a PhD thesis at the University of Birmingham under Professor L.A. Clark – another former colleague, whose research team at Birmingham has contributed a great deal to engineering aspects of deteriorology. This PhD thesis, linked to reports produced by former C&CA colleagues and by TRL, is the basis for the guidance given in Chapters 5 and 6.

Last but not least, there is the mechanics of producing the text. Jackie Fitch worked hard at making sense of my manuscript and translating it into a respectable typescript. Isabel Harvey reproduced the Figures, often from rough sketches. Katy Low at Taylor and Francis was of great help in moving the book into final production.

Chapter 1

Introduction

1.1 Background

As individuals, and probably without thinking in a systematic way, we routinely take decisions in our everyday lives, which could be classified as management and maintenance. To repair and maintain, or to replace by new, this might relate to personal things like shoes, household furniture and fittings, or the family car. Mostly, these are items which have relatively short lives, and we are spoilt for choice, with upgraded or more attractive versions appearing on the market. Increasingly, here, the trend is to buy new – the culture of the so-called 'throw-away society'.

Even in our homes, attitudes are changing. The traditional activity of regularly painting wooden windows and doors is being challenged by the advent of plastic, which apparently requires much less maintenance. Upgrading is becoming more common, either in the form of insulation or by building extensions, as family needs change, i.e. due to changes in our expectations or performance requirements.

As we move out into the wider world of infrastructure, the picture, at least so far, is rather different. Expectations are for longer lives (however ill-defined) but, initial design has not really taken service life or maintenance needs into account; design for the time factor has largely been via material and component specifications. It is not clear why this has been so in practice. Design lives for different types of structure have been given in Codes (in the UK) for over 50 years, but largely ignored. Over the same period, major attempts have been made to develop and rationalise maintenance regimes in a general way, the importance of which cannot be over-emphasised, since maintenance work was estimated at 30 per cent of the construction industry budget 40 years ago, and this has now risen to over 50 per cent.

For concrete structures in particular, the situation has been complicated by durability issues. While some types of aggressive action have been known for some time, e.g. sulfate attack, new forms have appeared, e.g. thaumasite. Physically, abrasion and frost attack have also been known, and attempts have been made to deal with these via specifications. We have also had the spectre of alkali–silica reaction (ASR), now well researched and understood. However, the major hazard has been corrosion, due either to carbonation of the concrete or to the effects of chlorides from various sources, and especially from the use of de-icing salts on roads.

In research terms, durability has become a growth industry, aimed mainly at understanding the mechanisms involved and the effects that they produce. The impact that these effects can have on structural performance and strength is less well defined; researchers in this area have tended to take a general approach, based on risk analysis and probabilistic methods.

While striving to understand and use all the new scientific information, owners have been faced with deteriorating structures and the need to manage these on a day to day basis, with function, safety and serviceability in mind. In general, the basis for rational decision-making has not always been clear. Is the priority to prevent, or at least slow down, the rate of deterioration or is it to repair, upgrade, strengthen, or rebuild? What is the most effective action, and, just as important, when should it be taken, in optimising the balance between whole life costing (WLC) and the maintenance of satisfactory technical performance?

In response to this need, the development of repair and preventative measures has also become a growth industry. Different categories of remedial action have evolved, based on different principles and with different objectives, and, within each category, a plethora of options are now available on the market. Based on development and laboratory testing programmes, there has been strong reliance on manufacturers' literature in the past, and it is only very recently that Standards have started to appear, particularly through CEN and EN 1504 (see Chapter 7). While there is still a dearth of reliable data on the long-term performance of repairs in the field, feedback is beginning to appear, presenting a perspective of perceived expectations and actual reality. As a result, the repair industry itself is becoming much better organised on a more scientific basis.

The importance of asset management and maintenance will become greater in the future. The ancient myth that concrete lasts for ever has proven to be false, due to a combination of unforeseen or disregarded aggressive actions and less-than-perfect design and construction. The resulting need to maintain structural performance has to be seen alongside the issues of functional obsolescence, rising expectations in performance requirements and increases in loadings. Thus, asset management has to become more hands-on and proactive, with much less recourse to the crisis management tools. In addition, sustainability will dictate the need to maximise the use of existing infrastructure.

In general, the need to think in terms of whole life performance (and costs) has gained acceptance, but how best to achieve that is much less clear. Owners of structures are having to cope as best they can, but progress has been made for certain types of structure at least; concrete bridges are a particular example, simply because they are the most threatened by de-icing salts. It is a dynamic period of change and evolution, and this book is targeted at providing an overall perspective – with the emphasis on a straightforward engineering approach.

1.2   The book in outline

1.2.1   General

As indicated in the Preface, the scope of the subject matter is vast, making it impossible to provide a full treatment of all aspects. The intention is to provide a perspective while concentrating on principles, practices and procedures; the extensive references provide the essential details.

The concept is to keep the approach as simple as possible, consistent with the quality and reliability of the inputs. Assessment of concrete structures is not, nor ever will be, an exact science. Risks have to be managed, but there will always be uncertainty, requiring engineering judgements to be made – whether in assessment itself or in evaluating a repair strategy – in setting margins for selected solutions. The book is also based on the concept of progressive assessment, i.e. in taking the investigation only as far as is necessary, to permit a decision on action to be taken with sufficient confidence. This is illustrated in Figure 1.1, a flow chart which is repeated in Chapters 3 and 4. The suggestion is that sufficient information may have been gathered at any one of the four levels, as the investigation proceeds. This investigation will almost always involve a mix of Overview and Insight, as indicated in Figure 1.2. Overview will always be necessary. Insight will depend on what information is readily available, and on the nature and severity of the deterioration.

Chapters 2–6 are closely related to Figure 1.1, and they offer an indication of the balance between Overview and Insight which may be necessary at each level. This culminates in Chapter 6 on detailed assessment (level 4 in Figure 1.1) and is followed by Chapter 7, which gives guidance on the selection of the most appropriate remedial measure.

Section 1.2.2 summarises the most important points made in each of these chapters.

1.2.2   Scope and key points

1.2.2.1   Chapter 2 Feedback and perspective on deterioration

Chapter 2 provides a brief summary of feedback on in-service performance, while identifying the relative importance of the main deterioration

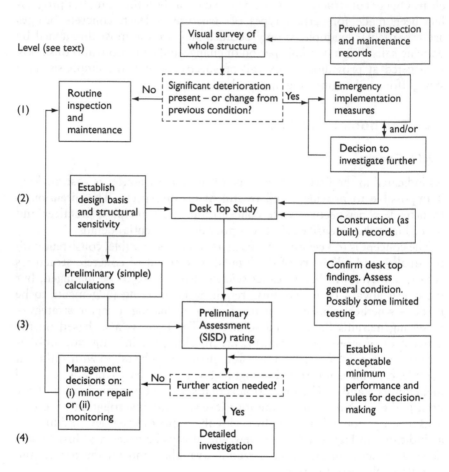

Level (see text)

(1)

(2)

(3)

(4)

Figure 1.1 Flow diagram for progressive assessment, showing four levels where a decision on action might be taken

mechanisms for different types of structure. Of equal importance are the primary causes of the deterioration, since there are lessons to be learned here, not only in giving focus to inspecting and assessing existing structures but also in making changes to current practice for new construction.

The general environment, and local micro-climate, is a major factor influencing the scale of deterioration. This is especially true regarding moisture, which acts as a transport mechanism for aggressive actions. Structures which have efficient joints and drainage systems perform better under all circumstances. This suggests that provision to improve the control of water should feature strongly in any planned remedial action. There is also clear evidence that deficiencies in detailing and execution are major contributors.

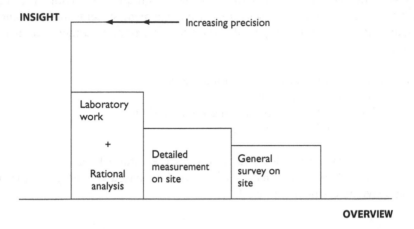

Figure 1.2 The essential balance between Overview and Insight, in assessment

Concrete cover is frequently highly variable, as is the quality of the concrete in different parts of the structure. This is a very clear message from feedback, and yet, in the 21st century, the main thrust, in striving for greater durability, is still in material specification terms.

In the relatively few cases where physical collapse has occurred due to deterioration, there are usually several contributing factors, not a single unique cause. Structural sensitivity requires careful consideration in assessment, and there is a need to be clear on what is acceptable in terms of minimum technical performance. On the other hand, many structures have hidden strengths, in terms of their ability to carry loads in ways not foreseen in the original design.

The contribution from this chapter is a perspective on past performance in service, which acts as a guide in establishing procedures in the chapters which follow.

1.2.2.2 Chapter 3 Management and maintenance systems

In the UK, major efforts to collect cost data on management and maintenance began in the mid-1950s. This exercise also shed light on technical issues which led to high costs of maintenance, but this never really got translated into practice for new construction. Beginning in the mid-1960s, this information was shaped into an asset management format, particularly for buildings, culminating more recently in a series of British and ISO Standards. Bridges were a world apart, as the ravages of de-icing salts became apparent, beginning in the 1970s; here, the main thrust had been a series of guidance documents issued by the Highways Agency.

Deteriorating concrete structures are not a unique UK problem alone, and this chapter presents a selection of management systems and guidance documents for different types of structure. This is done under four main headings:

(1) General national and international guidance
(2) Structure-specific guidance
(3) Guidance relating to specific aggressive actions
(4) Inspection and testing procedures.

From this, some principles and objectives are established, which underpin the procedures explained in Chapters 4–6, based on the principle of progressive assessment (Figure 1.1).

1.2.2.3 Chapter 4 Defects, deterioration mechanisms and diagnosis

Preliminary and detailed assessment procedures explained in Chapters 5 and 6 are focused on corrosion, freeze–thaw action and alkali–silica reaction. Other aggressive actions can occur and this chapter indicates what they are and where they are most likely to occur (Table 4.4).

The chapter starts from the broad church of defects generally, before homing in on the effects of deterioration mechanisms. It is crucial to detect the prime cause of the observed deterioration, since this will affect any prognosis on future rates and influence the nature of the remedial measures. Again, the influence of local micro-climate is shown to be critical.

Available test methods are identified and an indication given on how (and when) these are best used in the diagnostic phase. A key feature in all of this is the pre-determination of what is acceptable in terms of minimum performance as far as damage classification is concerned, since this will affect what tests are done and the depth of the investigation as a whole.

Basically, this chapter is about investigation and diagnosis, reviewing and utilising all the boxes in Figure 1.1 as necessary, to obtain the required input for the Preliminary Assessment state (level 3).

1.2.2.4 Chapter 5 Preliminary structural assessment

This chapter is founded on the formal numerical approach to damage classification, as proposed in the CONTECVET Manuals [5.1]. The CONTECVET project demonstrated that the basic approach, developed earlier for ASR, could be adapted for both corrosion and freeze–thaw action.

At this critical stage in assessment, all the information gleaned from the activities implicit in the boxes in the upper half of Figure 1.1 is being pulled together and evaluated. This requires a systematic approach, not least in collating and classifying the data from surveys and carefully selected tests. Interpretation of that data is then crucial. However, it also has to be put in context. The approach is a formalised version of damage classification (one of several possibilities given in Chapter 3), but it must be integrated with an engineering perspective of structural issues. This approach and philosophy is embodied in Chapter 5, and reflected in its sub-sections:

5.1   Introduction
5.2   Interpretation of test data
5.3   Engineering perspective in support of damage classification
5.4   Preliminary assessment: general principles and procedures
5.5   Preliminary assessment for ASR
5.6   Preliminary assessment for frost action
5.7   Preliminary assessment for corrosion
5.8   The nature and timing of intervention.

In the final Section 5.8, some guidance is given on relating the timing of an intervention to the values obtained for the structural severity rating. This is dealt with in more detail in Chapter 7 on remedial measures.

It should be recognised that the process is neither an exact science nor a straightforward linear progression. The data may be inadequate and some uncertainty may still exist, requiring further investigation. As a simple aid, it is helpful to put the severity rating into one of three groupings: Satisfactory; Borderline; Possibly Inadequate. This is illustrated in Table 1.1 for ASR. Action for groups 1 and 3 is relatively straightforward; group 2 may require further investigation. The most common follow-up action for groups 2 and 3 will be to carry out a detailed assessment (Chapter 6).

1.2.2.5   Chapter 6 Detailed structural assessment

The approach to detailed structural assessment made in Chapter 6 is essentially deterministic, using conventional design – related methods, modified by measured or assessed inputs for mechanical and section properties to represent the effects of deterioration. Some evaluation of risk is included, but more reliance is placed on the use of partial factors and a re-evaluation of minimum technical performance. The fact that fewer assumptions are made in assessment (e.g. on real loads and section properties, which can be measured) means that lower factors of safety might be used without affecting the overall reliability. Throughout, the design basis is Eurocode 2.

For simplicity, the analysis-of-structure stage is separated from the assessment of the residual strength of critical sections; this is because deterioration may affect element stiffness differently, compared with strength.

Table 1.1 Relating structural severity ratings to possible actions for ASR

Initial structural severity rating	Condition in the context of ASR	Action	Comments
N D	Satisfactory	Nothing beyond standard inspection routine	Easy decision
C B	Borderline	Conservative choice from: • Detailed assessment • Limited action • Monitoring • Load testing	Difficult area to decide on action, may need more investigation
A	Possibly inadequate	Remedial works and load testing depending on Detailed Assessment results	Relatively easy decision

Notes

a The rating values [SISDs] in column 1 are those in reference [3.39] for ASR and are indicative only here. Full details for all aggressive actions are given in Chapters 5 and 6.

b Most structures may contain elements with a combination of SISDs. Elements with similar SISDs should be grouped together for management purposes.

c The timescale for actions will depend on the structure, its condition, future use and life requirements. These can be broadly related to the SISD.

d Any strategy should be aimed at ensuring structural adequacy, 'increasing life' and improving appearance. This is likely to be conservative at this stage.

e It may be necessary to proceed to a detailed assessment before proceeding to the stage of deciding on management options.

Considerable space is devoted to evaluating the effects that deterioration might have on the different action effects such as bending, compression, shear, bond and anchorage – and on how this can be calculated using modified design equations. Performance in bending and compression is relatively easy to predict, shear and bond less so – and therefore receive more attention. In doing this, considerable reliance has to be placed on realistic experimental data. For ASR, the author is fortunate in having access to data banks made available by the University of Birmingham and the Transport Research Laboratory (TRL) – and by former colleagues of the British Cement Association. Corresponding data on the effects of frost action is very limited and this is reflected in the tentative nature of the recommendations made in Chapter 6.

For corrosion, the experimental data comes mainly from three sources – TRL and the University of Birmingham yet again, plus Geocisa in Spain as part of the CONTECVET programme. In analysing this data, the author is particularly indebted to a former colleague at the British Cement Association, M.P. Webster, whose PhD thesis [6.11] makes a major contribution in this field.

As with Chapter 5 on preliminary assessment, none of this is an exact science, nor is it ever likely to be. However, the principles appear to be sound, and further refinement will emerge as more data becomes available.

1.2.2.6 Chapter 7 Protection, prevention, repair, renovation and upgrading

This is a vast subject and a detailed treatment of all aspects is well beyond the scope of this book. Nevertheless, it is an integral part of asset management of concrete structures; coverage of the principles and processes involved is essential, while giving references for the detail.

A listing of the sections gives a flavour of the scope and approach:

7.1 Introduction
7.2 Performance requirements for repaired structures
7.3 Classification of protection, repair, renovation and upgrading options
7.4 Performance requirements for repair and remedial measures
7.5 Engineering specifications
7.6 Moving towards the selection process
7.7 Performance of repairs in service
7.8 Timing of an intervention
7.9 Selecting a repair option – general
7.10 The role of EN 1504 in selection
7.11 Selecting a repair option in practice
7.12 Concluding remarks.

In classifying the available options (Section 7.3), the EN 1504 system is adopted [7.13]. This covers all current possibilities and sets out 11 principles for repair, and the author believes that this document will shape the future for the repair industry. It is supported by a wealth of Standards (Appendices 1 and 2), and an attractive feature is the emphasis placed on execution and quality control while stressing the need for monitoring and after-care.

Before selecting a repair option in a particular case, it is important to be clear on what is required in performance terms, both for the repaired structure (Section 7.2) and for the chosen remedial option (Section 7.4). At all stages, the emphasis has to be on performance, since it is clear that the prescriptive approach followed in the past has led to cases where there were mismatches between what was expected and what could actually be delivered.

The selection process itself probably involves following a flow chart such as that shown in Figure 1.3. This is taken from the CONREPNET project [7.16], with the designated Principles (M1–M11) being those in EN 1504.

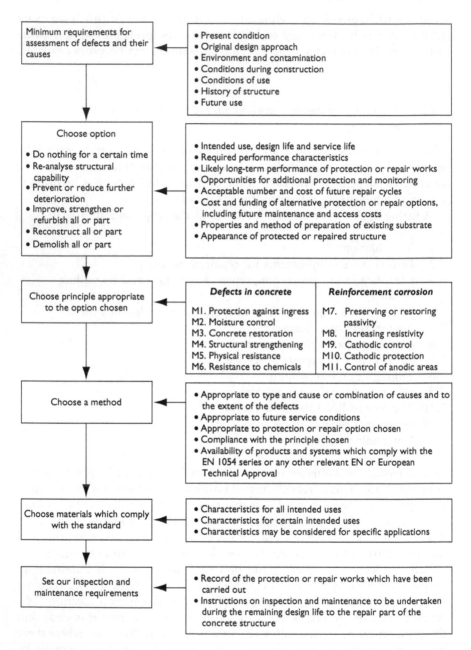

Minimum requirements for assessment of defects and their causes
- Present condition
- Original design approach
- Environment and contamination
- Conditions during construction
- Conditions of use
- History of structure
- Future use

Choose option
- Do nothing for a certain time
- Re-analyse structural capability
- Prevent or reduce further deterioration
- Improve, strengthen or refurbish all or part
- Reconstruct all or part
- Demolish all or part

- Intended use, design life and service life
- Required performance characteristics
- Likely long-term performance of protection or repair works
- Opportunities for additional protection and monitoring
- Acceptable number and cost of future repair cycles
- Cost and funding of alternative protection or repair options, including future maintenance and access costs
- Properties and method of preparation of existing substrate
- Appearance of protected or repaired structure

Choose principle appropriate to the option chosen

Defects in concrete	Reinforcement corrosion
M1. Protection against ingress	M7. Preserving or restoring passivity
M2. Moisture control	M8. Increasing resistivity
M3. Concrete restoration	M9. Cathodic control
M4. Structural strengthening	M10. Cathodic protection
M5. Physical resistance	M11. Control of anodic areas
M6. Resistance to chemicals	

Choose a method
- Appropriate to type and cause or combination of causes and to the extent of the defects
- Appropriate to future service conditions
- Appropriate to protection or repair option chosen
- Compliance with the principle chosen
- Availability of products and systems which comply with the EN 1054 series or any other relevant EN or European Technical Approval

Choose materials which comply with the standard
- Characteristics for all intended uses
- Characteristics for certain intended uses
- Characteristics may be considered for specific applications

Set our inspection and maintenance requirements
- Record of the protection or repair works which have been carried out
- Instructions on inspection and maintenance to be undertaken during the remaining design life to the repair part of the concrete structure

Figure 1.3 Overview of how the EN 1504 procedures might fit into the progressive assessment process (due to CONREPNET [7.16])

An important section in Chapter 7 is 7.7, which summarises the data obtained from the CONREPNET project on the performance of repairs in service. This gives a perspective, in terms of what might be expected based on development work and testing compared with what might be achievable on site; it again stresses the significance of the environment in affecting real performance. The timing of an intervention is also important in influencing the choice of remedial action. This is covered in Section 7.8, with figures suggesting a zonal approach, depending on the current position of the structure on the performance–time curve.

Moving onto the selection process itself, many current methods involve a numerical approach based on performance indicators and weighting factors. These can vary considerably in terms of complexity and level of detail. The one outlined in Section 7.9 stems from the CONREPNET work [7.16]. In the longer term, the author believes that this will be replaced by methods based more firmly on EN 1504, involving clearly defined performance requirements backed by test methods and quality-control systems.

Finally, Section 7.11 takes the reader through the selection process at a practical level, while broadly following the flow chart given in Figure 1.3.

1.2.2.7   Chapter 8 Back to the future

The heading for Chapter 8 is deliberate, in briefly using a crystal ball to look ahead, while extrapolating from past and present experience. The demand for asset management is likely to increase with sustainability adding momentum, and the current annual costs of repair variously quoted as 50 per cent of the output of the construction industry or in excess of £1 billion. One would like to think that lessons from past experience could somehow be translated into doing better in the future with new construction – not continuing to make the same mistakes.

There is unquestionably a greater awareness, and advances might reasonably be expected in all sectors represented by the chapters of this book. The advent of EN 1504 should add structure and quality to the repair process and already there are signs that the industry is responding. Testing technology is also expected to be a major force, with techniques improving beyond all recognition, especially the Non-destructive testing (NDT) methods. It is reasonable to expect, therefore, that continuous monitoring will feature more strongly, in support of normal inspection and management regimes. The evolution of computer-based systems will continue, permitting better control and interrogation, and giving owners more scope to develop alternative management strategies. A brave new world!

Chapter 2

Feedback and perspective on deterioration

In this chapter, a brief review is given of feedback on performance in service where defects have occurred in different types of structure. The relevance of doing that, in an asset management scenario, is to provide a perspective of the main causes of deterioration and of the variability that can occur. This is of benefit in taking decisions on the nature and scope of investigations which may be necessary to fully understand the present condition of an asset and to underpin assessment, prior to deciding on the nature and timing of any necessary action.

In 1986, the author published a paper entitled 'The design life of concrete structures' [2.1]. The main purpose was to review what was then known about durability and to make recommendations for improvements in the design and construction of concrete structures. The argument advanced was that the approach had to be holistic, embracing design and construction issues, in addition to the traditional approach of relying solely on material specifications. The evidence for that came from survey data available at that time. The sense of that data will be provided here, together with more recent information. The objective is to seek out trends and issues, which are of value in assessment.

2.1 General survey data

A broad-brush starting point is the data presented by Paterson [2.2] based on the analysis of 10 000 insurance cases in France; Paterson reported information of the type shown in Figure 2.1. This related to all common construction materials. The message is very clear; defects, and deterioration in general, would be greatly reduced if more attention was given to design and construction – together. These trends, with only minor differences in the percentages, have been confirmed in other countries, including the USA and Switzerland.

Paterson did not define 'defects', nor was it entirely clear what came under the headings of 'design' and 'construction'. Moving more firmly into the

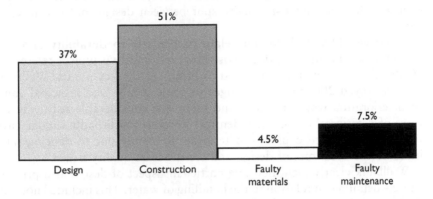

Figure 2.1 Causes of defects in terms of frequency of occurrence (10 000 cases, France: Paterson [2.2])

deterioration arena, other surveys – more in-depth and related to particular types of structures – can shed some light on this, while confirming the same general trends. Over the years, BRE has published a number of reports relating to deterioration and the condition of different building types. One such report relates to prefabricated, reinforced concrete houses, built to meet an urgent need in the post-war period up to the late 1950s [2.3].

This largely similar family of structures consisted of a number of different 'systems', which, with the benefit of hindsight, could have been improved conceptually. Concrete cover was low, and concrete technology did not promote good workmanship or long life; the use of calcium chloride as an accelerator was the norm – a practice subsequently banned because of its influence on reinforcement corrosion. The condition of these houses became an issue some 30–50 years after they were built. Detailed investigations revealed that the major problem was corrosion and that many houses were in need of repair although structurally sound. The rate of deterioration varied widely, and was expected to continue, although estimates of the time for cracking to occur were up to 30 years in some cases. While this family of structures were investigated as a group, they were subsequently managed on an individual basis, via a mix of routine maintenance and refurbishment. Collectively, they provide an interesting example of the causes of deterioration and of the options available in terms of management and maintenance, once an overall perspective is obtained.

Possibly the type of structure that has been studied most in the field is the concrete highway bridge. The reason is obvious: this type of structure is vulnerable to corrosion, particularly where de-icing salts have been used. Major condition surveys have been made, which are either general [2.4] or relating to particular bridge types [2.5, 2.6]. General surveys tend to show that the percentage of concrete bridges that are structurally deficient is relatively low. Nevertheless, particular surveys identify defects and deficiencies which are almost equally split between design and construction causes.

References [2.4–2.6] do not relate particularly to durability and corrosion. Such information does exist however, and two typical references, Wallbank and Brown, can be used to bring out the key points. Wallbank [2.7] surveyed 200 highway bridges in some detail. The principal cause of deterioration was corrosion, and there was considerable scatter in the rates of deterioration for near-identical structures in broadly similar environments. Identified causes of deterioration, mainly due to de-icing salts, involved the usual mix of design and construction quality issues.

Wallbank clearly identified one particular aspect of design as a primary cause – detailing, in relation to the handling of water. This included not only drainage and waterproofing, but also the type of joint, and how movement, including the possibility of cracking, was dealt with. Figure 2.2 illustrates this. Bridges classified as 'Poor' had a significant number of leaking joints; those as 'Good' had no joints. This is an important point in assessment, first in identifying the more sensitive structures and second in suggesting one possible solution in prolonging life, by introducing protective systems for better control of water.

The work by Brown [2.8] confirms that by Wallbank, in terms of basic findings and variability. He conducted general surveys of 92 bridges in three different locations, each having different local environments. From

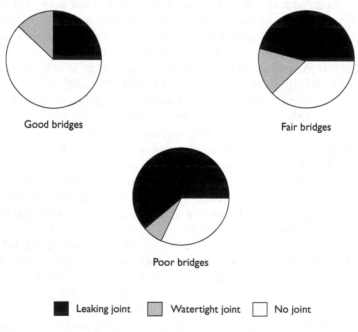

Good bridges   Fair bridges

Poor bridges

■ Leaking joint   ▨ Watertight joint   □ No joint

Figure 2.2 Influence of joints on bridge condition [2.7]

this pool, 17 bridges were selected for detailed investigation, including measurement of durability-related parameters such as oxygen permeability, capillary porosity and water absorption. The bridges were built between 1915 and 1972, and a particular aim was to study the influence of different cements, mix proportions and covers used over that period.

The more modern concretes were less susceptible to carbonation, but were used in bridges where location and function meant that they were subject to much more regular dosages of de-icing salts than the older structures. This was a major factor which determined condition. That condition, in turn, was influenced by the passage of water containing de-icing salt solutions, particularly where it drained from decks through joints onto cross-heads and the tops of columns; this was more influential than chlorides coming from traffic spray. Interpretation of the measurement of the durability-related parameters mentioned above was virtually impossible due to the high scatter; the results were compatible with a diffusion model, but did not define it closely.

Relating the findings from Wallbank [2.7] and Brown [2.8] to structural assessment, it is clear that:

- performance (nature and rate of deterioration) is highly variable;
- with older bridges, low covers and mix proportions may lead to corrosion due to carbonation, when dosages of de-icing salts are low or non-existent;
- chlorides, from whatever source, are the major risk, with the magnitude of their effects being highly dependent on local micro-climates. Local micro-climates are probably more critical when created by leaking joints or by cracks forming for whatever reason than by the general moisture conditions adjacent to 'sound' concrete;
- there is a strong inference that construction quality is a major factor, either in terms of low cover or in terms of concrete whose durability properties may not be as good as those deduced from strength measurements on standard control specimens during construction. This aspect will be looked at later in this chapter.

While the above statements are derived from survey data on bridges, there is reason to believe that the situation is similar for structures in maritime environments, in general [2.9].

To obtain an even broader perspective of the frequency and causes of defects and deterioration, a representative benchmark survey is required. However, this does not exist for the UK. What do exist are published surveys of individual or groups of structures, and attempts have been made to analyse these. One such review covered 271 cases [2.10], and, though possibly not being accurate in quantitative terms, does give some clear indications.

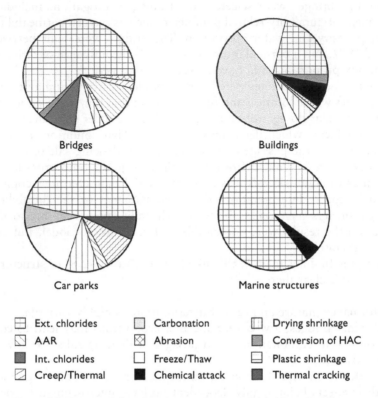

Bridges   Buildings

Car parks   Marine structures

⊟ Ext. chlorides	▨ Carbonation	▨ Drying shrinkage
◺ AAR	⊠ Abrasion	▨ Conversion of HAC
■ Int. chlorides	☐ Freeze/Thaw	⊟ Plastic shrinkage
◪ Creep/Thermal	■ Chemical attack	■ Thermal cracking

Figure 2.3 Reported deterioration mechanisms for bridges, buildings, car parks and marine structures [2.10]

The frequency of occurrence of the various mechanisms for deterioration for different types of structures is shown in Figure 2.3. For bridges, the most common cause of deterioration is corrosion due to external chlorides, whereas for buildings it is corrosion due to carbonation. Car parks are also prone to chloride-induced corrosion, with frost action being the next most significant. For marine structures, corrosion due to external chlorides again dominates, although this pie-chart can be misleading as the sample included a large number of maritime structures having dimensions similar to those for bridges (piers, etc.). For more massive structures directly exposed to wave action, abrasion was important, and there were a few cases of ASR. If the results in Figure 2.3 are combined to give an overall picture, then corrosion induced by external chlorides and corrosion due to carbonation, in that order, are the two most significant mechanisms, making up over 50 per cent of the cases. In assessment work, it is always important to identify the cause of the deterioration; it is not uncommon for more than

Table 2.1 Factors contributing to the failure of all structures reviewed in reference [2.10]

Factors contributing to deterioration	No. of cases	No. of cases (%)
Low cover	47	11.6
Environment	156	38.5
Poor quality concrete	64	15/8
Poor design detailing	29	7.2
Poor workmanship	17	4.2
Wrong specification	6	1.5
Failure of joint/waterproofing	31	7.7
Inadequate conceptual design	2	0.5
Wrong material selection	53	13
Total number of cases	405	100

one mechanism to be involved, and in such cases the most dominant one has to be identified at the investigative stage.

A further feature of this particular review [2.10] was the attempt made to identify the factors responsible for the deterioration. A summary of this is given in Table 2.1. Some difficulty was experienced in assembling this table, because of the different terminology used by the original authors. The structures were located in environments of varying aggressivity, and a cause attributed to 'Environment' can only mean that the structure could not resist the environment in which it was placed, in general terms. This is the biggest factor in Table 2.1, meaning that failure could not be attributed to any particular cause, but represented a failure in the system that produced the structure. Having said that, it is of interest to note the other listed factors. This is a mix of design, detailing and construction quality issues, indicating that the relevance of the findings by Wallbank [2.7] and Brown [2.8] may be fairly widespread.

2.2 Environment and local micro-climate

The general surveys outlined in Section 2.1 all indicated considerable scatter in performance, even for nominally similar structures in common environments. One possible reason for that is variability created by local micro-climates, and how these micro-climates react with the outer fabric of the structures.

It is well known that the role of water is important in durability terms, either as a transport medium for aggressive actions or as a key element in the deterioration mechanisms themselves. This has been well researched, and models have been developed to represent the processes involved; some of these are now used directly in design, generally for major structures in aggressive environments. However, the normal design process involves the specification of concrete cover and quality for defined environmental classes, which, over the years, have become more precise and detailed.

What is the relevance of this brief background to an assessment situation? First, there is a need to be aware of the nature and aggressivity of the local environment; this can affect possible future rates of deterioration. Second, design specifications are targeted at solid structural elements (beams, columns, etc. which are uncracked); there may be variations here in practice due to differing standards of workmanship (see Section 2.3), in terms of both cover and concrete quality, which may require consideration in assessment.

Local environments immediately adjacent to structures will largely be dominated by the combined effects of wind, temperature and water – depending also on both location and orientation. The possible impact that this might have, in durability terms, will depend on how that environment interacts with the outer fabric and boundaries of the structure itself. Movement, for whatever reason, can cause joints to open or cracks to form. Water, possibly driven by wind, can penetrate into these vulnerable areas, creating moisture conditions which can vary from vapour, through spray or driven rain, to run-off or ponding. These moisture conditions are almost unique to each structure, although these may again vary daily, seasonally or in the long term.

To understand the importance of this, it is necessary to look ahead and identify the factors which research has shown to be significant in influencing the magnitude and intensity of deterioration mechanisms. Table 2.2.

Table 2.2 Possible tabulation of environmental loads

CI Environmental loads due to carbonation
Actions

1) CO_2 concentration
2) Humidity–Temperature
3) Oxygen concentration at rebar surface
4) Chlorides

Influencing concrete properties
1) Porosity [permeability]
2) Binding capacity [Alkalis and CaO content]
3) Concrete humidity content

Climate classification
Micro actions
a) Wind direction
b) Condensation: sun orientation, temperature changes, closed spaces
c) Wet interiors
d) Leaks

Macro environment
a)   Dry or sheltered from rain
b)   Continuously wet
c)   Not sheltered from rain [wet-dry]
d)   Coated

C2 *Environmental loads due to chlorides*
Actions
1)   Chloride concentration. Type of cation
2)   Humidity–Temperature
3)   Oxygen concentration at rebar surface
4)   Carbonation [hydroxide content]

Influencing concrete properties
1)   Permeability [Diffusion coefficient]
2)   Binding capacity [C_3A content]
3)   Concrete humidity content
4)   Alkalinity [Cl^-/OH^- threshold]

Climate classification
   Micro actions
   a)   Wind direction
   b)   Wash out from rain
   c)   Condensation: sun orientation, temperature, closed spaces
   d)   Leaks–Joints
   e)   Irregular distribution of chlorides

   Macro environment
   a)   Marine
         i)   Air–sheltered, exposed
         ii)   Submerged
         iii) Tidal
         iv)   Splash
   b)   De-icing salts
         i)   low cycling [Kg NaCl/year], frequent cycling
   c)   Industrial

presents a brief summary of corrosion, due to either carbonation or chlorides. It may be seen that this represents a mix of concrete characteristics and climate classifications both at micro- and macro-level. Particularly, attention is drawn to micro actions in the table, since these can have a major influence in practice, and accurate input on these is essential in modelling the ingress of water, CO_2 or chlorides as shown in Figure 2.4.

This is by no means easy, and there is very little published information on how local micro-climates can vary over time, in practice. Nevertheless, the issues illustrated in Table 2.2 and Figure 2.4 require consideration

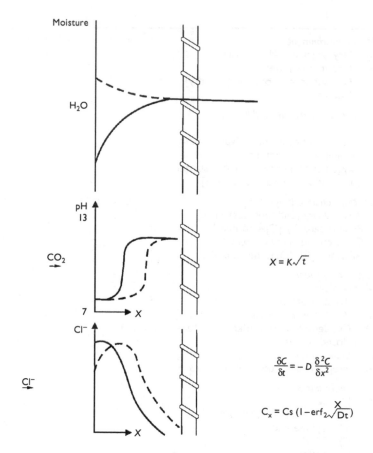

Figure 2.4 Figurative profiles of moisture, carbonation and chloride in the concrete

in individual cases, and assessments must be made, both in identifying vulnerable areas and in evaluating the most crucial micro-climate.

One type of structure where general progress has been made in identifying the most critical zones is the sea wall. This is illustrated in Figure 2.5, showing four different zones over the height of the wall. The reason for this is that the local micro-climates can vary, and this in turn can create different transport mechanisms by which chlorides penetrate into the concrete, as well as affect the internal conditions for corrosion to occur (availability of water and oxygen). Depending on ambient temperature, probably the most critical zones are (1) and (2).

For bridges, the identification of vulnerable areas and critical micro-climates is much more difficult because:

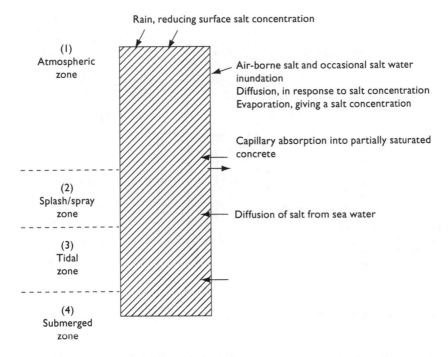

Rain, reducing surface salt concentration

(1)
Atmospheric
zone

Air-borne salt and occasional salt water
inundation
Diffusion, in response to salt concentration
Evaporation, giving a salt concentration

Capillary absorption into partially saturated
concrete

(2)
Splash/spray
zone

Diffusion of salt from sea water

(3)
Tidal
zone

(4)
Submerged
zone

Figure 2.5 Different critical zones for chloride-induced corrosion in sea walls

- exposure conditions can vary considerably from region to region, and
  even locally, depending on location and orientation; and
- the type of structure has a major influence in terms of:
  - overall design concept (e.g. determinate or indeterminate);
  - engineering detailing of outer boundaries;
  - provision for drainage;
  - efficiency of joints; and
  - general level of built-in protection.

Figure 2.6 shows some possible micro-climate conditions for a bridge subjected to de-icing salts. This identifies four different forms that water can
take, and what actually happens will depend very much on the behaviour
of the joints. The resulting moisture conditions will also depend on the
finish of the concrete surfaces and on whether or not cracks are present. It
is difficult to generalise here, but there is qualitative evidence that support
systems are especially vulnerable when ponding and rundown can occur.

Faced with visual evidence of deterioration in routine inspections, assessment engineers may consider more detailed investigation, possibly with

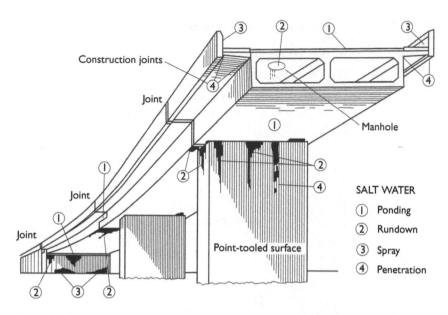

Figure 2.6 Various micro-climate conditions for a bridge subjected to de-icing salts

some intrusive testing. To get maximum benefit from this, it is important to study local micro-climate possibilities in the bridge as a whole, to ensure that the most critical areas are identified. One possible way of doing this is to identify and study the individual moisture movements one by one. Such an example is shown in Figure 2.7, when considering salt-bearing spray; this also clearly illustrates zones requiring special attention in designing for new construction [2.11].

Moving on to buildings, it should first be noted that there are certain types of buildings which are better classed as civil engineering structures in durability terms, either because they are open and exposed or because their function makes them vulnerable to specific aggressive actions. Sports structures, such as grandstands, or transport structures, such as bus stations, are included in the first category. Multi-storey car parks are in the second, due to risk from de-icing salts. There are also types of building which have to be treated as special cases, such as nuclear installations or chemical plants.

However, the vast majority of buildings are essentially enclosed structures. Facades may vary from fully exposed concrete to 'indoor' concrete frames totally enclosed by cladding. Micro-climates will be dictated by the factors shown in Figure 2.8, to a degree also controlled by location and orientation.

Relating Figure 2.8 to Table 2.2 and Figure 2.4, controlling features in durability terms will be the usual mix of concrete characteristics and local micro-climates. Again, the detailing of facades will be a major factor, in

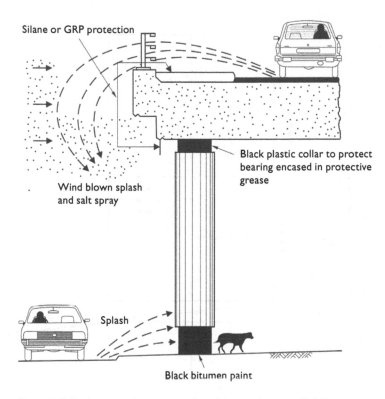

Figure 2.7 Bridge boundary areas vulnerable to salt spray [2.11]

terms of identifying critical areas. Here, it is necessary to be aware of the possible mechanisms by which rainwater can penetrate through joints. Some of these are shown in Figure 2.9.

The main purpose of writing this brief Section 2.2 is to give the assessment engineer a sense of perspective. Table 2.2 may represent a target, in terms of what he has to consider, but the precision of what is achievable, especially in terms of climate classification, is much influenced by practical issues – leading to the suggested strategy of treating each structure as unique. Getting to know as much about the structure as possible, carefully identifying the most critical zones and making sensible estimates and judgements about the most aggressive local micro-climate, as influenced by its contact with the structure itself, are the initial steps to be carried out by the assessment engineer. Without these, any subsequent analysis, no matter how sophisticated, may be widely out in predicting future deterioration rates.

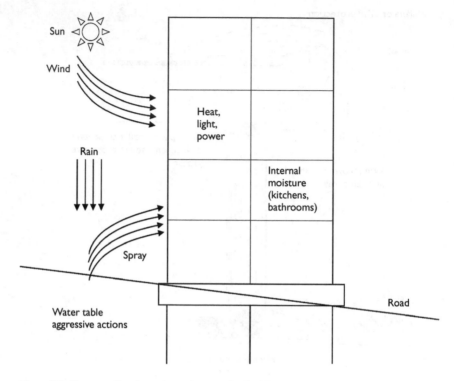

Figure 2.8 Factors affecting micro-climates for buildings

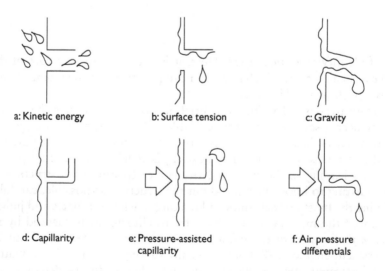

a: Kinetic energy   b: Surface tension   c: Gravity

d: Capillarity   e: Pressure-assisted   f: Air pressure
                       capillarity   differentials

Figure 2.9 Possible mechanisms by which rainwater leaks through joints

2.3   Effect of design, detailing and construction quality

The purpose of this section is to review feedback from site surveys on variations in the quality of the concrete and concrete cover that can occur in practice, due to varying standards in design, detailing and construction – relative to what might have been expected from the original design and with durability issues in mind. This is a preamble to Chapter 3, which deals with deterioration mechanisms – their diagnosis and likely effects. The focus is on concrete quality and cover, in relation to corrosion.

In 1985, Dewar [2.12] graphically described, as shown in Figure 2.10, the different types of 'Cretes', that can occur when casting a simple beam. Dewar's paper was about testing for durability, and the point he was making was that compliance was most commonly via strength measurements on standard control specimens (Labcrete). It is generally accepted that the strength of concrete in structural elements is less than that measured on standard control specimens; indeed for decades designers have taken this into account in strength calculations by introducing partial safety factors and other coefficients into their equations. How these reductions relate to the different zones shown in Figure 2.10 is not really known with any certainty, but are most probably related to Heartcrete, and the system seems to operate quite well as far as strength is concerned.

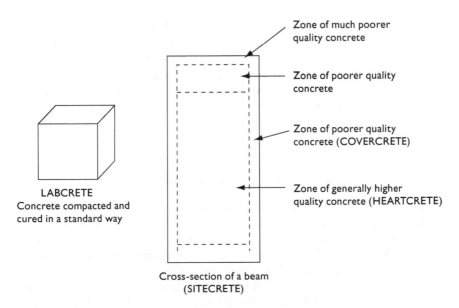

Figure 2.10 The various 'CRETES', according to Dewar [2.12]

For durability, however, the major concern is with Covercrete and the zones of poor quality concrete in Figure 2.10. Here, the concern is more with absorption and permeability characteristics (see Figure 2.4) than with strength. Most survey data do not provide direct information on these properties, but are likely to depend on workability, compaction and curing, in minimising the formation of capillary channels, and water/cement gradients as placing proceeds. To get a feel of concrete variability in practice, it is necessary to rely on the measurement of strength, as determined from cores or by a range of NDT methods. The translation of such measurements into assessments of desirable durability characteristics has to be done with some care, since they are not linearly related to strength. If such a relationship can be established, then a picture of in situ variability can emerge from the field surveys.

Most of the field data in the UK were obtained prior to the early 1990s, and the results obtained may be modified for more recent construction and technology, e.g. self-compacting concrete and the use of admixtures. Focus on this subject was provided by the University of Liverpool and Queens University Belfast, who both had a strong interest in developing new NDT methods for measuring in situ strength. The brief summary which follows is centred on their activities ([2.13] and [2.14], which also give access to data from other sources).

Strength variations within an element depend on the type of element. Based on extensive field measurements, Bungey et al. [2.13] also found that strength decreased with height in vertically cast elements. As an example, they produced Figure 2.11, giving ranges of strength at the top and bottom of different types of element, compared with what might be expected from standard control specimens. Reference [2.13] also showed the possible existence of strength contours as shown typically in Figure 2.12. for a beam and a wall. Again, there is a geometrical factor at work, so also an effect due to levels of compaction (operator skill and care). Other available data show variations in these general patterns due to different percentages of reinforcement.

Typical in situ equivalent 28-day cube strength

The evidence shown here suggests that there is a geometrical factor involved, linked to higher water/cement ratios as height increases, with compaction being particularly difficult very close to the top (Figure 2.10). Part of the problem is the lack of a definition of what constitutes adequate compaction and a means of knowing when it has been achieved; specifications tend to be of the type 'the concrete shall be adequately compacted'. Studies of this are thin on the ground, but some guidance is available [2.15]; this links vibration effort to chloride permeability and freeze–thaw damage.

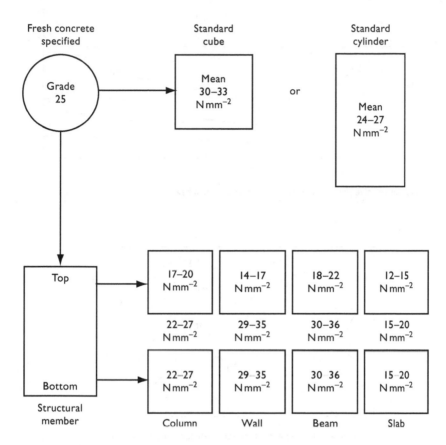

Figure 2.11 Typical relationships between standard specimen and in situ strengths [2.13]

Another important factor, not included in the above review, is the influence of curing, which is often on the critical path in construction terms and possibly not always given the attention it deserves. Its importance in durability terms has been known for some time [2.16–2.18], particularly for cover zones, in promoting a dense impermeable concrete with good durability characteristics in relation to concrete strength.

In any detailed investigation into the current condition of an existing structure, the above evidence suggests the need to study the quality of the concrete in the cover zone, using methods which can be related to absorption and permeability characteristics. Just as important is the need to establish the thickness of that cover.

There is evidence on cover thickness from field surveys of a wide range of structural types in different countries over a period of 30 years. References [2.19–2.25] give some examples. Covers are recorded in terms of mean, maximum and minimum values, together with standard deviations. The

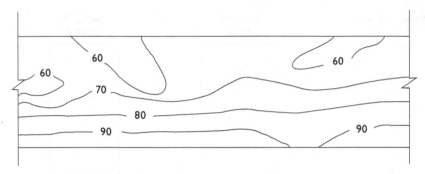

(a) Typical relative percentage strength contours for a beam

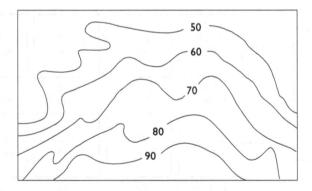

(b) Typical relative percentage strength contours for a wall

Figure 2.12 Typical in situ strength contours [2.13]

picture that emerges is basically the same. While there were some differences for different types of element, there was also great variability within each. Mean values were generally slightly less than the specified nominal cover, but there were cases of zero cover and of maximum values twice the nominal value.

Clark [2.25] described the failure to achieve specified cover as both widespread and chronic. His work included identifying the occupational origin of the defects. This showed a fairly even split between site operatives and what he called 'management'; this latter group included designer/detailers as well as site management, emphasising the impor-tance of buildability issues in the achievement of durable structures. This is reflected in more recent guidance [2.26], covering equally both tech-nical and organisational issues, with the latter including co-ordination,

Table 2.3 Concrete cover requirements in UK Codes (mm)

Code and year		All external concretes	Grade 40 concrete Exposure classes			
			Mild	Moderate	Severe	Very Severe
Building Codes						
DSIR	1933	12				
CP114	1948	25				
CP114	1957	38				
CP114	1969	40				
CP110	1972		15	25	30	60
BS8110	1985		20	30	40	50
ENV 1992.1.1 (Eurocode 2)			15	25	40	40
Bridge Codes						
BS5400	1984			30	35	50
BS5400	1995			40	45	60

communication, checklists and compliance matters – most of which do not formally exist, except for post-tensioning systems [2.27].

In assessment terms, the implication is that more recent structures may well have been built to higher standards, but the need to be aware of the causes of possible defects is important in any assessment. An awareness of how specified values for concrete cover have changed over the years is also relevant. This is shown in Table 2.3 for the UK. In general, values have increased over the years, with the first attempt made in 1972 to classify the environment in simple terms. This trend is reflected in other countries, although the actual values can vary considerably.

Much of this has been consolidated in the final version of Eurocode 2, [2.28] supported by National Application Documents and material standards [2.29]. This involves more detailed definitions of environment classes and a greater interplay between specified values for concrete quality and cover for each of these.

In summary, for assessment, while the latest design standards and specifications have brought improvements, variability in concrete quality and cover is a factor, whose significance becomes more important when considered with the influence of detailing (Section 2.1) and of micro-climate and its interaction with the outer fabric of the structure (Section 2.2).

2.4 Performance of remedial/repair systems in the field

Quantified feedback on the performance in service of repair/protective systems is sketchy. There are numerous individual case studies, involving damage assessment and giving reasons why a particular remedial action was selected. Some of these have been collected together [2.10, 2.30]; while these are interesting to read, it is difficult to detect any general guidance or even a common denominator for assessing alternative options. Owner's attitude would appear to be important, as indeed is cost, in the absence of quantified data.

In response to a perceived growth market, very many repair and remedial systems have evolved, and this process is still ongoing, with new formulations and techniques being introduced. Much of this has been via proprietary materials and systems, mostly supported by laboratory research, and it is only recently that any sort of practical perspective is beginning to emerge, e.g. [2.31] on patch repairs. In parallel, consensus guidance documents began to appear on how to carry out different types of repair – e.g. in the UK by The Concrete Society [2.32–2.34], in North America by the American Concrete Institute [2.35] and internationally by *fib* [2.36]. There are numerous other examples.

These guidance documents were based both on research data and on practical experience in implementing repairs. It is apparent that the standard of workmanship is crucial for an effective repair, in terms of both pre-preparation of surfaces and the installation process itself. Other lessons that have been learned (but not always applied) include:

– the need for compatibility between parent concrete and repair material in terms of strength and stiffness [2.31];
– the importance of good bond;
– the need for electro-chemical compatibility between concrete and the repair material. Frequently, the precise nature of the corrosion process is not well understood, and the extent of the repair area is less than it should have been ([2.37] and [2.38]); and
– the need to see budgets and costs, which are crucial, alongside predicted technical performances [2.39].

The process briefly described above is evolutionary, and what is really required is quantified data from the field. A recent case study is of interest [2.40]. This involved a bridge in an aggressive environment where a number of different repair and protective methods had been applied, with the effectiveness of these being assessed some 12 years after their installation. Their effectiveness was variable, with the dominant influence being the quality of preparation and installation.

A qualitative survey of 70 case histories of bridges from 11 European countries has been provided by Tilly [2.41] as part of CONREPNET, an ongoing EU-funded project. The bridges had been built between 1908 and the 1990s, with some repairs in place for up to 33 years. The results from the study of bridges were also compared with some 140 case histories of

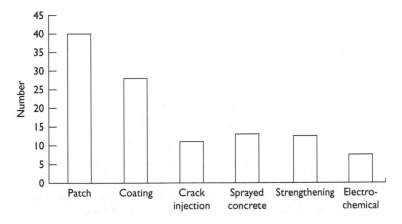

Figure 2.13 Types of repair – Tilly (Reproduced from reference [2.41], with permission from the author)

other structures, including buildings, dams, car parks, power stations and industrial structures. The repair systems included:

– patch repairs (cementitious or polymer modified);
– coatings;
– crack injection;
– sprayed concrete;
– electro-chemical; and
– restoration of strength.

The numbers of each type of repair are shown in Figure 2.13.

In classifying the performance of the repairs, a three-level system was used:

	Class	
(1)	successful; no signs of deterioration;	45 per cent
(2)	some evidence of deterioration (e.g. minor cracking or discolouration of coatings), but not necessarily requiring remedial action;	25 per cent
(3)	failure, clearly requiring remedial action.	30 per cent

For the bridges, the percentages in each class were as shown above. For the other structures, the picture was similar, with fewer failures (22 per cent) and more successes (53 per cent).

Figure 2.14 breaks this general data down to show the classification for different ages of repair. For up to 5 years, 80 per cent of the repairs were

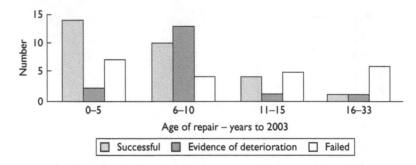

Figure 2.14 Classification of performance versus life of repair – Tilly (Reproduced from reference [2.41], with permission from the author)

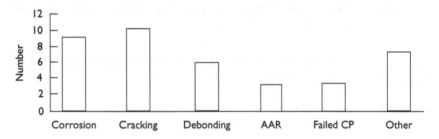

Figure 2.15 Reported modes of repair failures – Tilly (CP refers to cathodic protection) (Reproduced from reference [2.41], with permission from the author)

satisfactory; for up to 10 years, 30 per cent were satisfactory and for up to 25 years, the 'satisfactory' figure was 5 per cent. The 'other structures' category reflected these figures closely.

The principal modes of repair failure are shown in Figure 2.15. Continued corrosion, cracking and debonding applied to both patch repairs and coatings.

Attempts were also made to relate failures to identifiable causes; the results are shown in Figure 2.16. In some cases, more than one cause was identified, with incorrect design of the repair and method of application being associated with poor workmanship.

The value of the case studies presented by Tilly stems from the fact that they relate to performance in practice rather than in controlled laboratory conditions. They also cover the linkage between diagnosing the nature and extent of the deterioration to the repair process itself, including its subsequent inspection.

The data in Figures 2.13–2.16 show considerable scatter, with 20 per cent of the repairs failing in the first 5 years and some still being effective after 33 years. It is difficult to deduce any generally acceptable life

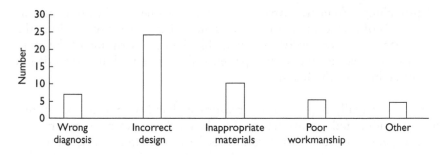

Figure 2.16 Relating failures to identifiable causes – Tilly (Reproduced from reference [2.41], with permission from the author)

expectancy – something that an owner would be looking for, when taking a decision to adopt a particular repair option. This suggests that available guides, while improving all the time, are either not being applied properly or are still deficient in some respects. Based on his findings, Tilly [2.41] had the following suggestions to make:

- more attention to investigating and diagnosing the original cause of the deterioration – its nature and its extent;
- independent checking of the complete design process for the repair, to ensure both aptness and compatibility;
- close supervision of the repair work in the field;
- greater use of NDT both in acceptance testing and in subsequent inspections;
- better training for operatives, especially in design, ensuring that the processes are properly understood.

As will be seen later in Chapter 7, major efforts are being made to define and standardise different classes of repair systems and materials, while introducing supporting test methods. Inevitably, this is largely based on controlled laboratory work. In due time, that approach should link up with feedback from performance in service. Section 2.4 is introduced into this book, to make the user aware of the need for that link.

2.5 Structural issues

2.5.1 Historical context

Hammond [2.42] quotes some interesting remarks attributed to Robert Stevenson in 1856, when, as president of the Institution of Civil Engineers, part of his summing up after the presentation of a paper contained the following:

[–]nothing is so instructive to the younger members of the profession, as records of accidents in larger works, and of the means of repairing the damage. This was really more valuable than a description of the more successful works, and such experience should be faithfully recorded in the archives of the Institution.

This has not really happened, for reasons of liability or whatever. While some individual failures are reported in detail, the numbers are insufficient to draw general conclusions, certainly as far as the effects of deterioration are concerned. Some historical records of early experience do exist however [2.42–2.43].

Feld [2.44] is interesting in the present context, since he classified failures under the following headings:

1. Design deficiencies
2. Construction problems
3. Durability of concrete
4. Foundation problems.

Feld is mainly reporting on American experience, and, over 40 years ago, durability was already emerging as a separate entity, with de-icing salts and corrosion, frost action, chemical attack, abrasion and reactive aggregates all being identified. In general, his few examples of actual failures reveal a previous lack of design awareness of aggressive actions and of the effects that these can have on structural concrete.

The few failures recorded by Feld [2.44] were predominantly due to lack of durability and involved pre-stressed concrete tanks, reservoirs and pipes located in aggressive environments. Failure in these cases involved physical collapse. In these extreme cases, there are generally several underlying causes.

Three more recent collapses underline this point. The first is the Ynys-y-Gwas bridge in Wales, which collapsed under self weight only [2.45]. This bridge consisted of a number of longitudinal I-beams, formed by post-tensioned short lengths of precast beam. The deck was also post-tensioned transversely through the flanges. There were, therefore, a large number of joints filled with mortar, with only cardboard formers protecting the tendons through the joints. The deck did not have a concrete overlay. The bridge was over a river in a valley, with de-icing salts able to penetrate into the joints through the porous mortar. The result was corrosion of the tendons, leading to the failure.

The second example also relates to a bridge, this time in Canada, as briefly reported in *New Civil Engineer* [2.46]. The collapse occurred in late morning, when the bridge was in use, and unfortunately led to a number of fatalities. The collapsed section of the bridge is shown in Figure 2.17 (a) and an elevation of the failed region in Figure 2.17 (b), with the major failure plane identified.

The underlying cause was reinforcement corrosion, due to de-icing salts. The salts penetrated the movement joint at the half-joint at the end of the cantilever unit. The precise reinforcement detailing in the half-joint is not known, but it is a notoriously difficult area to reinforce effectively – for that reason, this type of joint is no longer used for bridges in the UK. What we are seeing here is sensitive structural detail, linked to deficiencies in the movement joint, thus permitting the ingress of the de-icing salts to a critical local bearing area. Actual collapse requires a number of underlying causes, and certainly structural sensitivity requires careful scrutiny when aggressive media are involved.

(a) Plan view of the collapsed section
(b) Elevation of the failed region (cantilever and suspended span combination) showing the probable failure plane.

Figure 2.17 Bridge failure in Canada (2006) [2.46]

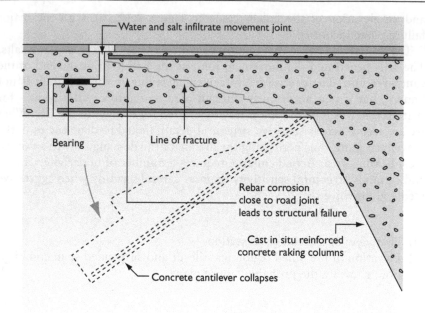

Figure 2.17 (Continued)

The third case study relates to the Pipers Row four-storey car park in Wolverhampton [2.47]. This was a flat-slab structure, built by the lift-slab method, whereby the floor slabs are cast on the ground and jacked into position, being supported on wedges cast into the previously erected columns. This form of construction is sensitive to erection procedures and to tolerances in the positioning of the wedges on the four sides of each column. The structure was built in 1965 in accordance with the then prevailing Code, CP 114:1957.

At approximately 3 a.m. on 20 March 1997, part of the top deck of the car park collapsed, under its self weight only. Figure 2.18 shows a plan of the car park and the location of the failure area, which spread over approximately 6 bays. A view of the failure area is given in Figure 2.19, where it can be clearly seen that the principal mode of failure was punching shear.

An extensive physical and theoretical forensic investigation was undertaken at the request of the Health and Safety Executive.

The conclusions which emerged from this study included the following:

- the basic failure mode was punching shear, probably initiated at an interior column, with only self weight to support;
- the as-built resistance to punching shear was adequate, assuming no deterioration in the slab;

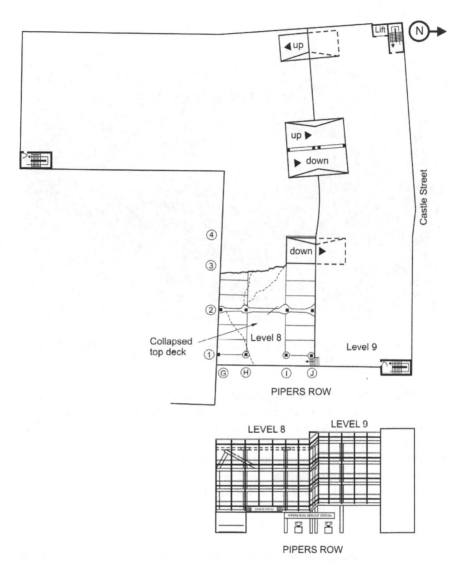

Figure 2.18 Plan of Pipers Row car park showing the extent and location of the collapsed area (Courtesy of the Health and Safety Executive [2.47])

- punching shear resistance was eroded by deterioration of the concrete in the top surface of the slab close to the columns. The breadth and depth of this advancing deterioration affected the bond and anchorage of the top reinforcement, thus altering and reducing the combination of internal forces which normally provide resistance to punching shear.

Figure 2.19 Collapsed top floor slab at Pipers Row car park, Wolverhampton on 25 March 1997 (Courtesy of the Health and Safety Executive [2.47])

Other factors such as temperature or the uneven distribution of load around the columns may have contributed to the failure, but the dominant factor was deterioration of the concrete in the top of the slab close to the columns;

- the quality of the concrete in the top slab was variable and not as good as that in the other floors;
- water frequently ponded on the open top-slab deck, and the water-proofing system had failed locally and progressively, resulting in the slab being saturated for much of its life, and therefore vulnerable to frost attack;
- the friable concrete that had previously been observed close to the columns was consistent with surface scaling and internal mechanical damage typical of frost action. There was no significant corrosion of any of the reinforcement;
- examination of samples of the debris after the collapse showed that deterioration of the concrete had penetrated to the top reinforcement. Theoretical calculations showed that this depth of deterioration would reduce the once-adequate resistance of the underteriorated as-built slab to a level corresponding to the imposed forces acting at the time of collapse.

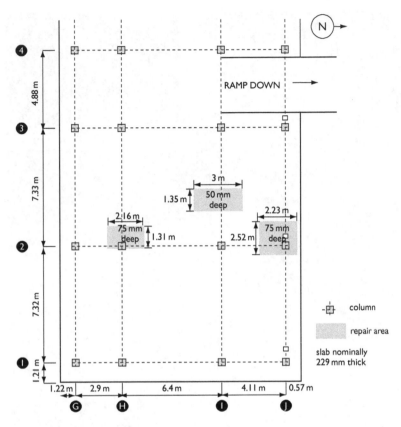

Figure 2.20 Plan of repaired areas in collapse zone at Pipers Row (Courtesy of the Health and Safety Executive [2.47])

Pipers Row was an interesting forensic study, leading to fairly definite conclusions. Frost action was the dominant factor, in this case leading to failure, caused by more than one deficiency in the structure itself – poor quality concrete, a failed waterproofing system, and a structure that was inherently vulnerable to progressive collapse.

An interesting by-product to the Pipers Row case was the role of repairs. From the mid-1980s, concern had been growing about leakage and seepage of water, and attempts were being made to rectify this. Also, the expanding areas of friable concrete had been noted, and a decision was taken in early 1996, to make local repairs in structurally sensitive areas. The location of these is shown in Figure 2.20. Examination of the debris after the collapse showed that these were ineffective structurally, since they become detached from the parent concrete (Figure 2.21).

T₂ bar corrosion

T₁ bar corrosion

Figure 2.21 Column H2, showing the detached repaired area (Courtesy of the Health
and Safety Executive [2.47])

There are other cases in the literature where some type of aggressive
action has played a major role in the failure. The underlying message in
all of them is basically the same as that described in the above examples –
several factors, acting in combination, are involved, affecting the nature

and magnitude of any resulting deterioration. It is important in assessment work to recognize this fact and to try and identify what these are, since it will affect the nature and timing of remedial action.

2.5.2 What constitutes 'failure'?

A review of case studies does not reveal any consistent pattern in decision-making on what constitutes minimum acceptable technical performance, requiring remedial action of some sort. Possibly, this is not surprising, because of the complexity of the inter-related factors which affect such a decision. Some of these are subjective. Some are non-technical, including:

- future plans for the structure, including possible changes in use or increases in performance standards;
- management and maintenance strategy, e.g. a policy of preventative maintenance;
- budgetary constraints;
- owner's attitude to serviceability, functional or appearance deficiencies.

Having said that, some basic requirements do emerge.

Safety is of paramount concern; this includes strength, stability and structural serviceability. Here, decisions on when to act have tended to be conservative, to avoid or minimise the risk of physical failure, as outlined in Section 2.5.1. To some extent, this conservatism is now reducing, as knowledge and confidence increase, and greater reliance is placed on sequential analyses of increasing rigour, with input which is measured rather than assumed.

As far as deficiencies due to deterioration are concerned, there are two major issues:

1. Reduction in structural capacity, compared with that provided by the original design (bending, shear, bond, cracking, etc.).
2. The possibility of alternative modes of failure, due uniquely to the effects of deterioration.

On (1), the general approach on assessment will be analytical, or will involve the re-evaluation of critically loaded sections using modified design models – although there is a growing tendency to use reliability methods. In either case, it is necessary to derive input parameters which represent the physical damage due to the deterioration. In general, this includes changes in geometry or reductions in mechanical properties, which will depend on the nature and intensity of the deterioration mechanism. Feedback indicates that these need to be looked at closely in individual cases, to ensure confidence in the predictions of residual capacity.

Included in the above are obvious features such as loss of concrete or rebar cross-section, or cracking and spalling leading to reductions in bond or anchorage. Actions such as frost can produce surface scaling (reducing effective cover) or internal mechanical damage (reducing concrete strength and stiffness). The major effect of actions such as ASR is expansion, and the structural implications depend on the presence or otherwise of restraints.

In extreme cases of deterioration, issue (2) above can come into play. An obvious example is the spalling of cover concrete in a short column, which transforms it into a slender one. Local areas of stress concentration, such as bearings, may also be vulnerable, particularly if the reinforcement detailing is poor and the concrete is not contained by stirrups or links. This latter example raises a general point – the need to consider not only the strength, stiffness and stability of the structure as a whole, but also those of local regions that may be vulnerable in structural terms.

These aspects will be considered further in Chapter 6, when dealing with assessment. Here, the basic message from feedback is the need not only to develop and use analytical methods where input is representative of observed/measured deterioration, but also to develop a rationale for what represents minimum acceptable technical performance. This is crucial not only in settling on the most appropriate remedial measure but also in the timing of it.

If 'failure' is used in a general sense, then other aspects of minimum technical performance need to be considered. Mostly, these are at the serviceability level. Loss of function may be a major issue, due, say, to loss of stiffness or excessive deflections. Cracking, for whatever reason, may be a concern where containment is important, or the avoidance of leakage. Excessive cracking caused by corrosion along the length of the reinforcement can introduce additional concerns. If this leads to spalling of concrete on the corners of beams or the soffits of slabs, a safety issue, quite separate from strength reduction considerations, may arise due to falling debris.

2.5.3   Structural sensitivity versus hidden strengths

A further factor to be considered in assessing the urgency of intervention is the type and nature of the structure. Some structures are inherently more vulnerable to the effects of deterioration. The degree of redundancy is important here, plus the existence or otherwise of alternative load-carrying mechanisms – or even the quality of the reinforcement detailing. In the case of Pipers Row (see Section 2.5.1), the extent of the progressive collapse could have been limited had there been effective bottom reinforcement in the slab passing over the columns. The same type of thinking can be extended to individual elements or critical sections: Are anchorage or bearing zones confined by links? Are the elements over- or under-reinforced?

Is the deterioration located at critical sections, or at lap lengths? Does loss of stiffness locally affect the distribution of the load effects in the structure as a whole? And so on.

A pointer in the other direction is the existence of hidden strengths, not considered in the original design. A guidance document on this aspect has been published by the Concrete Bridge Development Group [2.48]. Quite apart from considering alternative load paths or forms of resistance, specific examples are given which include:

- compressive membrane action in slabs, in the presence of edge restraints;
- the use of the variable truss analogy in shear assessment;
- the benefits of transverse pressure from supports in providing confinement.

The emergence of this type of guidance is linked to a trend towards the progressive use of more rigorous analytical methods which better reflect the real structural capacity compared with that based on the generally conservative rules in Codes. For highway bridges in the UK, this approach was effectively formalised with the issuing of Advice Note B79/98 [2.49].

BA79/98 is based on the principle that assessment will begin at a simple level and progress further if necessary. Five levels of assessment are recognized:

Level 1 Assessment using simple analysis and codified requirements and methods
Level 2 Assessment using more refined analysis
Level 3 Assessment using better estimates of bridge load and resistance values
Level 4 Assessment using specific target reliability
Level 5 Assessment using a full probabilistic reliability analysis

Levels 1 to 3 are consistent with the satisfactory levels of safety which are the base for current semi-probabilistic methods contained in design Codes. This represents the norm for most cases, with reasonable efforts being made to represent the as-built structure modified by the effects of deterioration. If possible, consideration of structural sensitivity should be integrated into this approach.

Also, levels 1 to 3 have the advantage of giving a direct comparison with what was provided in the original design. Levels 4 and 5 come more into play when a decision has been made to consider a change in the overall reliability index required.

2.6 Summary of Chapter 2

Chapter 2 is intended to provide a perspective of asset management, and structural assessment in particular. This will invariably involve some form of analysis, and it is important that the input to that is realistic and representative. It is also important to be aware and to think positively in engineering terms.

Key features emerging from this feedback are as follows:

Section 2.1 General survey data

(a) The existence and extent of variability.
(b) While aggressive actions are the origin of deterioration, the magnitude of the effects will be strongly influenced by design, detailing and construction quality.

Section 2.2 Environment and local micro-climate

(a) The general environment – water, wind and temperature – has an influence on the scale of deterioration (Table 2.2).
(b) Much more important is the local micro-climate immediately adjacent to the outer surfaces of the structure. There is an interaction here with the structure itself, which will affect the moisture conditions inside the concrete and hence the creation of transport mechanisms (Figures 2.5–2.8).
(c) The efficiency of joints and drainage, and the existence of any cracks will exacerbate the impact of micro-climates.
(d) In deciding on what action to take to prolong the life of a deteriorating structure, consideration should be given to controlling the local micro-climate.

Section 2.3 Effect of design, detailing and construction quality

(a) Concrete quality will vary within a structure for reasons of geometry and workmanship.
(b) Cover to reinforcement and prestressing steel will also vary.
(c) There is a need to study the thickness and quality of the concrete in the cover zones, with the emphasis on durability parameters such as absorption and permeability.

Section 2.4 Performance of remedial/repair systems in the field

(a) There is lack of quantitative data of real performance under real conditions in the field, compared to that under controlled conditions in the laboratory.
(b) Experience in the field, though limited, indicates considerable variability (Figures 2.13–2.16), with strong influences due to:

- wrong diagnosis;
- incorrect design and method of application;
- workmanship, including pre-preparation and site practice.

Section 2.5 Structural issues

(a) Cases of physical collapse, where deterioration has dominated, are relatively rare. Where they have occurred, there are usually several causes (Section 2.5.1).
(b) In any structural assessment, it is important to have a clear idea of what constitutes minimum acceptable technical performance. This affects both the nature and timing of any remedial action (Section 2.5.2).
(c) Trends in structural assessment are moving towards progressive structural analysis, e.g. levels 1–5 in Section 2.5.3. In following this path, it is important to:

- derive realistic and representative input for the assessment parameters,
- be fully aware of structural sensitivity,
- take advantage, wherever possible, of any hidden strengths in the as-built structure.

References

2.1 Somerville G. The design life of concrete structures. *The Structural Engineer*. Vol. 64A. No. 2. February 1986. pp. 60–71. IStructE, London.

2.2 Paterson A.C. The structural engineer in context (Presidential Address). *The Structural Engineer*. Vol. 62A. No. 11. November 1984. pp. 335–342. IStructE, London.

2.3 Building Research Establishment (BRE). The structural condition of prefabricated reinforced concrete houses designed before 1960. *BRE Information Paper IP10/84*. BRE, Garston. 1984.

2.4 Dunker K.F. and Rabbat B.G. Performance of prestressed concrete highway bridges in the United States – the first 40 years. *Journal of the Precast/Prestressed Concrete Institute (PCI)*. Vol. 37. No. 3. May/June 1992. pp. 48–64.

2.5 Muller J.M. *25 years of concrete segmental bridges – survey of behaviour and maintenance costs*. Company Report. J. Muller International, San Diego, CA, USA. 1990. p. 48.

2.6 Podolny W. The cause of cracking in post-tensioned concrete box girder bridges and retrofit procedures. *Journal of the Precast/Prestressed Concrete Institute (PCI)*. March/April 1985. pp. 82–139.

2.7 Wallbank E.J. *Performance of concrete in bridges: A survey of 200 highway bridges*. A report prepared for the Department of Transport by G. Maunsell & Partners. HMSO, London. 1989.

2.8 Brown J.H. *The performance of concrete in practice: A field study of highway bridges*. Contractor Report 43. Transport Research Laboratory (TRL), Crowthorne, Berks, UK. 1987. p. 61.

2.9 Di Maio A.A. et al. Chloride profiles and diffusion coefficients in structures located in marine environments. *Structural Concrete*. Vol. 5. No. 1. March 2004. pp. 1–4. Thomas Telford Ltd., London.

2.10 British Cement Association. *Development of an holistic approach to ensure the durability of new concrete construction*. Final report to the Department of the Environment on Project 38/13/21(cc 1031). BCA, Camberley, UK. October 1997. p. 81.

2.11 Pritchard B.P. *Bridge design for economy and durability*. Thomas Telford Ltd., London, UK. 1992.

2.12 Dewar J.D. Testing concrete for durability (Part 1). *Concrete*. Vol. 19. No. 6. June 1985. pp. 40–41. The Concrete Society, Camberley, UK.

2.13 Bungey J.H., Millard S.G. and Grantham M. *Testing of concrete in structures*. 4th Edition. Taylor and Francis, London, UK. 2006.

2.14 Murray A. and Long A.E. A study of the insitu variability of concrete using the pull-off method. *Proceedings of the Institution of Civil Engineers*. Part 2. No. 53. December 1987. pp. 731–745. ICE, London.

2.15 Forssblad L. and Sallstrom S. Concrete vibration – what's adequate? *Concrete International*. September 1995. pp. 42–48. American Concrete Institute (ACI), Detroit, MI, USA.

2.16 Parrott L.J. Effects of changes in UK cements upon strength and recommended curing times. *Concrete*. Vol. 19. No. 9. September 1985. The Concrete Society, Camberley, UK.

2.17 Ho D.S.W. et al. The influence of humidity and curing time on the quality of concrete. *Cement and Concrete Research*. Vol. 19. No. 3. 1989. pp. 457–464.

2.18 Gowripalan N. et al. Effect of curing on durability. *Concrete International*. February 1990. pp. 47–54. American Concrete Institute (ACI), Detroit, MI, USA.

2.19 Johansson A. and Warris B. *Deviations in the location of reinforcement*. Report No. 40. Swedish Cement and Concrete Research Institute, Stockholm, Sweden. 1969.

2.20 Mirza S.A. and Macgregor J.G. Variations in dimensions of reinforced concrete members. *Proceedings of the American Society of Civil Engineers (ASCE)*. Vol. 105. No. ST14. April 1979.

2.21 Morgan P.R. et al. How accurately can reinforcing steel be placed? Field tolerance measurement compared to Codes. *Concrete International*. Vol. 4. No. 16. ACI, Detroit, USA. October, 1982.

2.22 CUR (Commissie Voot Uitvoering van Research, Holland). *Concrete Cover*. Report 113 (2 volumes) 1984. Available from Concrete Information Services, Camberley, UK.

2.23 Marosszeky M. and Chew M.Y.L. Site investigation of reinforcement placement in buildings and bridges. *Concrete International*. Vol. 12. No. 4. ACI, Detroit, MI, USA. April, 1990.

2.24 Clear C.A. *Review of cover achieved on site*. BRE Contract No. EMC/94/86. BRE, Garston, UK. December, 1990.

2.25 Clark L.A. et al. How can we get the cover we need? *The Structural Engineer*. Vol. 75. No. 17. pp. 29–296. 1997. Institution of Structural Engineers, London. UK.

2.26 CIRIA. *Specifying, detailing and achieving cover to reinforcement*. Funders Report RP561/4. CIRIA, London 1999.

2.27 The Concrete Society. *Durable bonded post-tensioned concrete bridges*. Technical Report 47 (2nd edition), 2002. The Concrete Society, Camberley, UK.

2.28 British Standards Institution. Eurocode 2: Design of concrete structures – Part 1.1: General rules and rules for buildings. BS EN 1992-1-1: 2004. BSI, London, 2004.

2.29 British Standards Institution. BS 8500; Concrete – Complementary British Standard to BS EN 206-1. BSI, London. 2002.

2.30 Mallett G.P. *Repair of concrete bridges*. State of the art review. Thomas Telford Ltd., London, UK. 1994. p. 194.

2.31 Canisius T.D.G. and Waleed N. Concrete patch repairs under propped and unpropped implementation. *Proceedings for the Institution of Civil Engineers*. Structures and Buildings 157. April 2004. Issue SB2, pp. 149–156. ICE, London, UK.

2.32 The Concrete Society. *The repair of concrete damaged by reinforcement corrosion.* Technical Report 26. 1984. The Concrete Society, Camberley, UK.

2.33 The Concrete Society. *Guide to surface treatments for protection and enhancement of concrete.* Technical Report 50. 1997. The Concrete Society, Camberley, UK.

2.34 The Concrete Society. *Diagnosis of deterioration in concrete structures.* Technical Report 54. 2000. The Concrete Society, Camberley, UK.

2.35 American Concrete Institute (ACI). *Concrete Repair Manual – 2nd Edition.* 2004. ACI, Detroit, USA. Available in the UK from BRE, The Concrete Society or the Concrete Repair Association.

2.36 International Federation for Structural Concrete (*fib*). *Management, maintenance and strengthening of concrete structures.* Bulletin 17. 2002. *fib*, Lausanne, Switzerland.

2.37 Barkey D.P. Corrosion of steel reinforcement adjacent to structural repairs. *ACI Materials Journal.* July–August 2004. pp. 266–272. ACI, Detroit, MI, USA.

2.38 Vaysburd A.M. and Emmons P.H. Corrosion inhibitors and other protective systems in concrete repair: Concepts or misconcepts. *Cement & Concrete Composites 26.* 2004. pp. 255–263. Elsevier Ltd., USA.

2.39 Yunovitch M. and Thompson N.G. Corrosion of highway bridges: Economic impact and control methodologies. *Concrete International.* January 2003. pp. 52–57. ACI, Detroit, USA.

2.40 Andrews-Phaedonos F. et al. Rehabilitation and monitoring of Sawtells Inlet Bridge – 12 years later. *Concrete in Australia.* Vol. 30. No. 3. September–November 2004. pp. 17–23. Concrete Institute of Australia, Sydney.

2.41 Tilly G.P. Performance of repairs to concrete bridges. *Proceedings of the Institution of Civil Engineers, Bridge Engineering.* Vol. 157. September 2004. pp. 171–174. ICE, London, UK.

2.42 Hammond R. *Engineering structural failures.* Odhams Press Ltd., London. 1956. p. 207.

2.43 Champion S. *Failure and repair of concrete structures.* Contractors Record Ltd., London. 1961. p. 199.

2.44 Feld J. *Lessons from failures of concrete structures.* American Concrete Institute (ACI). Monograph No. 1. 3rd printing 1967. p. 179.

2.45 Woodward R.J. and Williams F.W. Collapse of Ynys-y-Gwas bridge, West Glamorgan. *Proceedings of the Institution of Civil Engineers. Part 1: Design and Construction.* Vol. 84. August 1988. ICE, London.

2.46 New Civil Engineer (NCE). *Poor maintenance blamed for Montreal collapse.* NCE, London, UK. 5 October 2006. p. 5.

2.47 Health and Safety Executive (HSE). *Interim results of Pipers Row investigation.* News Release 30 April 1997. Also, www.hse.gov.uk, for details of the forensic investigation.

2.48 Concrete Bridge Development Group (CBDG). *Notes for guidance on the assessment of concrete bridges.* Technical Report No. 9. CBDG, Camberley, UK. 2005.

2.49 Highway Agency. *The management of sub-standard highway bridges.* BA 79/98. Highways Agency, London, UK. 1998.

Chapter 3

Management and maintenance systems

3.1 Introduction and history

In this chapter, a brief review is given of different types of management and maintenance systems and of how these have evolved to meet owner needs, both technical and non-technical. One objective is to bring out differences between systems; while overall purpose is broadly the same in all cases, differences do emerge due to owner's polices and strategies, and the nature, function and type of structure.

The scope of this book is limited to concrete structures. In the UK, the vast majority of concrete structures have been built in the last 50–60 years, and therefore the historical review is limited to that period, during which attempts were being made to collect data on maintenance costs and to introduce a more systematic and rational approach to the management of the infrastructure.

A useful starting point here is the paper by O'Brien [3.1]. With a strong emphasis on buildings, this provides a perspective on three issues:

1. A lament that the attempt made in 1950 by issuing a Code of Practice [3.2], which set out categories of life for both buildings and components, had not been successful, either in terms of influencing design or in having an impact on in-service management and maintenance regimes.
2. Feedback on the causes leading to excessive maintenance costs, which had shown that design/detailing and buildability were just as important as the longevity of the material components themselves.
3. A summary of the extensive work that had been done in the decade 1955–1965, in assembling data on maintenance costs (with the then Ministry of Public Building and Works and the Building Research Establishment central to this). This showed increased costs with the age of the structure and considerable differences between different components. There was also great variability generally, deriving from different attitudes and standards among owners in terms of the levels of

deterioration which they would tolerate before intervention. Strategies varied from systematic preventative maintenance to crisis management, frequently dictated by the availability of funding.

In an attempt to improve the situation, O'Brien [3.1] provided a review of what he called 'Building Economics' (now known as Life Cycle Cost Analysis, LCCA). He also gave strong support to the concept of a maintenance manual for individual buildings analogous to that for a car, while noting that one such manual had been published the previous year [3.3].

Subsequent developments over the next two decades saw the collection of even more cost data, but also a sharper focus on shaping this information into an asset management format. Still on buildings, reference [3.4] is typical.

While this approach quoted initial life periods for the structure and its main components, its major contribution was to isolate out individual elements and services in great detail (e.g. external joinery, sealants, rainwater pipes, tiles, boilers, etc.) and to give recommendations for each on:

- replacement life;
- maintenance-free life; and
- items to be considered in management and maintenance terms.

This automatically led to suggestions for periods between inspections, while defining different types of inspection as follows:

- Routine: continuous regular observations by the building user.
- General: annual visual inspections of the main building elements under the supervision of suitably qualified personnel.
- Detailed: full inspection of the building fabric by suitably qualified personnel at intervals not exceeding 5 years.

We thus see the introduction of suggested routines into the management regime and a strong suggestion regarding the experience and expertise of the people involved.

Progress in the last two decades has been towards codifying these procedures in British or ISO Standards. In terms of management, maintenance and assessment, there are BS 8210 [3.5] and ISO 13822 [3.6]. To assess full progress in codification, it is necessary to look at these alongside their counterparts on durability and service life. These are BS 7743 [3.7] and ISO 15686–1 [3.8].

It is uncertain at this time, how widely these documents are used as the basis of asset management systems in practice. However, since they relate to the scope of this book, and since their influence is likely to increase as new and more detailed versions appear, brief summaries are presented in Table 3.1.

Table 3.1 Brief summary of references [3.5–3.8], with respect to management and assessment

Code reference	Subject/Title	Scope	Comments/Special features
BS 8210 : 1986 [3.5]	Building maintenance management	Buildings General guidance Building fabric Engineering services	Largely in the form of a series of checklists, with little or no quantification. Some useful Appendices, with an emphasis on documentation and on standardization in recording data on condition and defects.
ISO 13822 : 2001 [3.6]	Assessment of existing structures	General framework of assessment: – Procedures – Preliminary assessment – Detailed assessment – Action and the environment – Structural analysis – Verification	Basically concerned with defining elements/stages in the assessment framework. Little quantification except in Informative Appendices, covering:- – flowcharts – evaluation of data from inspections in reliability terms – target reliability levels
BS 7543 : 1992 [3.7]	Durability of buildings and building elements, products and components	Definition of durability in design terms. Suggested lives for buildings and components. Recommended levels of maintenance Agents which can lead to deterioration (+ examples)	Both are concerned with new construction and have many common features; hence they are considered together here, with the emphasis on aspects which relate to management, maintenance and assessment. Defined lives for components are of help in setting intervals between inspection, as is the classification of these components as: Repairable; Replaceable; Lifelong. Maintenance levels are defined in reference [3.7], and the concept of categories of obsolescence in reference [3.8].
BS ISO 15686-1 : 2000 [3.8]	Service life planning Part 1 : General Principles	As above, but additionally, – forecasting – a factorial approach to design – obsolescence, flexibility and re-use – performance requirements and LCCA	There is some treatment of minimum acceptable performance as far as function and serviceability are concerned, plus 8 categories under the heading 'Effects of failure'.

The brief history presented so far, up to Table 3.1 could be classified as 'evolutionary supply', with focus on general guidance and buildings. Even the earliest references [3.1, 3.2] recognised the existence of aggressive actions, without giving much detailed guidance in terms of what to do about their effects within an asset management system. In parallel to that, a demand grew for solutions to be found, as awareness of the range and intensities of aggressive actions increased. As an example, the extensive use of de-icing salts in the UK did not occur until around 1960, and it was more than a decade before the scale of the effects was appreciated.

These 'solutions' came in many guises, nationally and internationally, and from different origins. If we classify what has been described so far (Table 3.1) as:

Type 1 *National and international guidance*, which is general but with emphasis on buildings (rather like design codes), the following additional types occur:

Type 2 *Structure specific*. Mostly, these originate with major owners or groups of owners. Examples are:

- bridges and highway structures;
- car parks (multi-storey and underground);
- ports and maritime structures;
- heavy civil engineering (dams, etc.); and
- nuclear installations.

Type 3 *Aggressive actions*. This type of guidance is generally produced by expert committees, with specific coverage of:

- alkali–silica reaction (ASR)
- corrosion;
- freeze–thaw attack; and
- sulfate attack or other related actions such as delayed ettringite formation or thaumasite.

Type 4 *Inspection and testing procedures*. This type relates only to part of the management and assessment process. However, it is an important part and has been isolated out here, since separate guidance does exist on it, to augment those given in the first three types.

The approach for each type can be very different, when dealing with deterioration. For Type 2, the tactic is generally to start with existing asset management systems and to explore how the latest information on deteriorology can be integrated into these. For Type 3, the emphasis is more on giving a full treatment of the particular mechanism – of its nature and how to avoid it – rather than on the detailed implications for asset management and maintenance. Where asset management is treated, the guidance given can be different in character, with different schools devoted either to a probabilistic approach or to using modifications to Code-based design equations.

The purpose of the remaining sections in this chapter is to give examples of each type of guidance, and to draw from these general principles and objectives, prior to setting out recommended procedures in Chapters 4–6.

3.2 Some examples of current guidance on asset management

This section is not intended to be comprehensive but to create awareness of what is happening in different sectors. The chosen examples are also intended to bring out key features, which will be fed into Section 3.3 on general principles and objectives. The sequence of presentation will follow the four types outlined in Section 3.1.

3.2.1 Type 1: General national and international guidance

3.2.1.1 Institution of Structural Engineers, UK. Appraisal of existing structures [3.9]

First published in 1980, and with a fascinating foreword presenting a brief, telling historical perspective of assessment, this publication covers all types of structure in a general way. It is referenced here for two main reasons:

1. The comprehensiveness of the flow charts, which make a good basis for similar, but simpler, diagrams, when focusing more on deterioration.
2. The approach adopted for considering the load combinations for analysis during assessment. This is based on partial factors, consistent with the semi-probabilistic approach in most limit state design codes. Because of the greater precision in known values for material properties and actual loads, reduced values are proposed for the partial factors, while maintaining the same reliability levels implicit in the original design.

3.2.1.2 American Concrete Institute (ACI). Strength evaluation of existing concrete buildings [3.10]

ACI Committee 437 first produced a report on this subject in 1967, and there have been several updates since. The basis for assessment is the design code ACI 318, with modifications introduced to account for measured or observed geometrical and mechanical properties. The progressive nature of inspection and assessment is clearly indicated, with a strong emphasis on the importance of records and documentation. Structural analysis is seen as being central to assessment, involving possible benefits from a more rigorous approach.

3.2.1.3 Canadian Standards Association. Existing bridge evaluation [3.11]

Although strictly applicable only to bridges, this reference is included here since the approach could be general, and it is an example of attaching a supplement to a design code, to deal with assessment.

An interesting feature is the approach to determining safety levels. While the methodology of evaluation is linked to the parent design Code, the required level of safety is established by going back to basics, in terms of target reliability index, β. The process is illustrated in Table 3.2.

Three factors affect the target values for β. These are:

(a) System behaviour, whose three categories are defined as:

   S1, where element behaviour leads to total collapse
   S2, where element behaviour probably will not lead to total collapse
   S3, where element behaviour leads to local failure only

(b) Element behaviour, whose three categories are defined as:

   E1, where loss of capacity is sudden, with little or no warning
   E2, similar to E1, but where retention of some post-failure capacity is likely
   E3, where failure is likely to be gradual, and warning of failure probable

(c) Inspection level, whose three categories are defined as:

   INSP1, where a component is not inspectable
   INSP2, where inspection is largely routine (as defined in the Code)
   INSP3, where detailed inspection is carried out on critical or sub-standard elements, and the results evaluated by calculation

The approach is interesting in terms of going back to basics (reliability indices) and linking this to a modified Code approach, which takes account of both the consequences of failure and its likely nature – all while making some allowance for what is known about the element in the field. The range of β values is quite wide, and requires good engineering judgement – something that is emphasised strongly in the Code.

3.2.1.4 CEB/fib activity [3.12, 3.13]

Fib was formed in 1998 by merging CEB and FIP. CEB, in particular, had a distinguished track record on matters relating to durability and service-life design, publishing numerous bulletins that reflect most of the activities in Europe in these subject areas. The bibliographies in both publications are of value in their own right.

Table 3.2 Illustration of Canadian Bridge Code approach to reliability [3.11]

System behaviour	Element behaviour	Target Reliability Index, β		
		Inspection level		
		INSP1	INSP2	INSP3
S1	E1	3.75	3.5	3.5
	E2	3.5	3.25	3.0
	E3	3.25	3.0	2.75
S2	E1	3.5	3.25	3.25
	E2	3.25	3.0	2.75
	E3	3.0	2.75	2.5
S3	E1	3.25	3.0	3.0
	E2	3.0	2.75	2.5
	E3	2.75	2.5	2.25

The two reports are considered together, since the scope is broadly the same, with reference [3.13] providing more of an overview and reference [3.12] operating at a more detailed level and in quantitative terms. This is reflected in Figures 3.1 and 3.2, flow charts taken from each report. Flow diagrams are introduced in this section to give some idea of current thinking and to act as a lead-in to later chapters in this book.

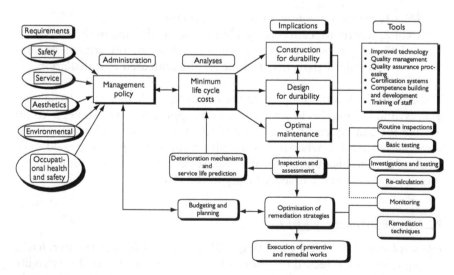

Figure 3.1 Relationship between requirements, minimum life cycle costs, and management and maintenance activities (Reproduced by permission of the International Federation for Structural Concrete [3.13])

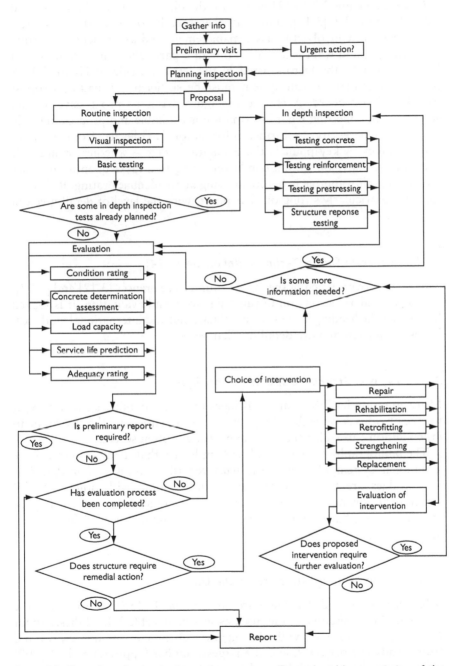

Figure 3.2 Flow chart for inspection and assessment (Reproduced by permission of the International Federation for Structural Concrete [3.12])

In effect, Figure 3.2 [3.12] is a more detailed version of the right-hand side of Figure 3.1 [3.13], and may therefore be of more direct interest to those actively involved in asset management and assessment. Nevertheless, strategy, and what is actually done in a particular timescale may well be controlled by the factors listed in the left-hand side of Figure 3.1. A positive asset management system must be sufficiently broad and robust to take this on board. A feature of Figure 3.2 is its progressive nature, starting with collecting what information is available and progressing only as far as is necessary, as further information is collected from the indicated sources. A key point is the evaluation stage – whether or not it is necessary to move on from a condition rating to a more quantitative analytical/reliability approach, before arriving at an adequacy rating. Reference [3.12] is particularly strong on that point, with the level of detail given in its Annex A.

3.2.1.5   Concrete Society Technical Report 54 [3.14]

This report covers much of the same ground as references [3.12] and [3.13], but is particularly strong on testing, diagnosis and evaluation. It is logged here under the heading Type 1: General national and international guidance, but is considered in more detail in Chapter 4.

3.2.2   Type 2. Structure specific – bridges

More attention has been paid to bridges than to any other structural type, for the obvious reasons that the populations are large, and the environmental conditions, to which these are exposed, are often severe.

The bibliography to reference [3.13] indicates that many countries have a national system of some sort, with regard to bridges. For that reason, bridges are treated separately in this section, with other types of structure following in Section 3.2.3. The coverage is not intended to be comprehensive, but examples are chosen to bring out some differences in approach.

3.2.2.1   Highway bridge structures in the UK

It has been estimated that there are approximately 155 000 bridges in the UK, with the breakdown of the stock as shown in Table 3.3. Of these, about 60 000 are in concrete, with the biggest percentage for motorways and trunk roads (80 per cent), and the largest number (approximately 43 000) being under the responsibility of local authorities.

A highway bridge assessment programme was initiated in the early 1980s and expected to last 14 years (up to 31 December 1998). The primary

Table 3.3 Breakdown of UK bridge stock

Type	Number
Motorway	5000
Trunk road	8000
Local authority	129 000
Railway	12 000
British Waterways	1000
Total	155 000

objective was to permit the passage of heavier vehicles with increased axle loads, while assessing as accurately as possible the current load-carrying capacity, including the effects of any deterioration.

The initial emphasis was on motorways and trunk roads and the basis for the work was provided by the Highways Agency, in a series of Standards and Advice Notes contained with the Design Manual for Roads and Bridges (DMRB). These have regularly been updated, and the website should be consulted for revisions and updates [3.15]. The approach in this documentation, and many of the detailed procedures, has permeated into the processes being followed for the other types of ownership shown in Table 3.3.

Some of the earlier guidance documents were brought out to address specific subjects, e.g. ASR or post-tensioned pre-stressed concrete. As the portfolio grew and became more coherent, it was acknowledged that the assessment of the bridge stock would become a continuous process, with a stronger emphasis on safety and whole life performance. This prompted a shift away from condition assessment and annual maintenance budgets towards a more systematic asset management approach, involving risk analysis, which maintained transparent minimum acceptable performance requirements at lowest whole life cost.

The basis for this overall coherence, and the change in the objectives, has been published [3.16] and the evolution of the system reviewed [3.17]. Emerging from experience and feedback, guidance is also available on recommended levels of maintenance spending [3.18] and the 'official' system is augmented by consensus guidance on particular aspects of the whole process [3.14, 3.19, 3.20]. A Code of Practice has also been published [3.21], which, while consistent with references [3.15–3.20, 3.22], gives a more strategic view of asset management, as may be seen from the contents list given in Table 3.4.

The evolution of practice in the UK for highway bridges over the past 20–30 years is probably not very different in principle from that in other countries. On the one hand, there was the need to cope with individual deterioration mechanisms such as corrosion or ASR; on the other hand, a programme was necessary to assess the capability of carrying

Table 3.4 Contents list for the UK Code on Management of Highway Bridges [3.22]

Executive summary – provides an overview of the code and key recommendations.

1. **Introduction** – describes the purpose, status and scope.
2. **Management context** – describes the overall management context and environment in which bridge managers have to operate.
3. **Asset management planning** – provides an introduction to asset management, and describes how it should be used for the management of highway structures.
4. **Financial planning and resource accounting** – provides an introduction to financial issues that may not be familiar to all bridge managers.
5. **Maintenance planning and management** – describes a process for developing and implementing cost-effective and sustainable maintenance plans.
6. **Inspection testing and monitoring** – describes regimes and techniques required to support good management practices.
7. **Assessment of structures** – describes a regime for structural review and reassessment and provides guidance on the assessment process.
8. **Management of abnormal loads** – provides guidance on the approaches that should be adopted for managing abnormal load movements.
9. **Asset information management** – describes an information management process, and the associated information requirements for highway structures.
10. **Framework for a bridge management system** – provides guidance on the functionality required by a BMS, in order to support full implementation of the code.
11. **Implementation of the code** – provides guidance on how bridge managers should approach the implementation of the code.

Appendices – there are 14 appendices which provide supplementary information.

increased live loads. The empirical methodology has gradually developed into a systematic asset management approach which is both strategic and detailed [3.21, 3.22].

Stripping the system down to its basic form produces the simple example shown in Figure 3.3, for two different bridges which have been assessed at time t_1. Although simple, the figure nevertheless illustrates a number of key features in the system, as follows:

1. Performance level on the vertical axis can be used at two different stages in the process:

    (a) at the preliminary assessment stage, when condition rating may be used to decide whether or not a more detailed analytical (numerate) assessment is necessary to check on the structural capacity;
    (b) at the detailed assessment stage, where performance, in terms of strength, stiffness, stability and serviceability, is the issue.

2. The system works only if the assessor has a clear picture of what is required in terms of critical (minimum) performance level, where, if reached, some action is necessary.

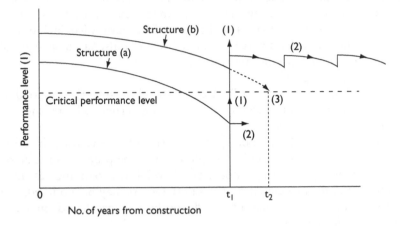

Structure (a)

Option	Maintenance strategy
1	Strengthen the structure now
2	Weight restrict structure/implement interim measures and monitor

Structure (b)

Option	Maintenance strategy
1	Strengthen the structure now
2	Undertake minor repairs to abate deterioration/monitor the structure
3	Do minimum work now and replace at reduced life t_2

Figure 3.3 Simple example of asset management for UK bridges

3. Establishing the gradient of the performance–time curve is important and can dictate the maximum interval between inspections.

4. In evaluating the performance–time curve at a detailed level, the central Advice Note in the system BA 79/98 [3.23] advocates 'the choice of the appropriate method of analysis will depend upon the structural form of the bridge and the required degree of accuracy. The simple methods, although conservative, are quick to use and should be tried initially where appropriate, before progressing to more accurate but more complex methods'.

In support of that philosophy, five levels of assessment are recognised:

Level 1 – Simple
The use of simple analytical methods known to give conservative estimates of load capacity.
Level 2 – Refined

More refined analysis and better structural idealisation. This may involve grillage analysis or finite element methods. Non-linear and plastic methods of analysis may also be used. This level also includes the determination of characteristic strength of materials, based on available data, but not on specific site testing.

Level 3 – Bridge specific

This would involve the use of Bridge Specific 'Assessment' Live Loading, in association with material characteristics determined by testing.

Level 4 – Modified criteria

Levels 1–3 are based on Code specified levels of safety, with the corresponding reliability related by implication to the previous satisfactory performance of the bridge stock. For Level 4, the load and resistance criteria and the associated partial safety factors may be modified. This may be done by a reliability study or by changes based on judgement/previous experience.

Level 5 – Full reliability analysis

Effectively, this requires probability data for all the variables defined in the loading and resistance equations. It is considered that this level is likely to be worthwhile only in exceptional cases.

The implication in all of these is that the greater rigour and precision required in moving through the levels will keep the performance–time curve in Figure 3.3 from reaching the critical level for a longer period of time. The strategy is to move towards a structure-specific situation.

3.2.2.2   BRIME, PONTIS and DANBRO

BRIME [3.24] was an EU-funded project targeted at studying bridge management systems in general and then developing a framework for the management of bridges on the European network, including the identification of the inputs needed to implement such a system. The study covered substantial computer-based systems, and two such systems, which were examined in some detail, were PONTIS from the United States and DANBRO from Denmark. Both are summarised in the Manual, a final deliverable from another EU-funded project REHABCON [3.25]. PONTIS was developed in 1989 for the Federal Highways Authority and is currently licensed to over 40 Departments of Transportation via the American Association of State Highway and Transportation Officials (AASHTO). DANBRO was originally developed for Denmark, but is also used by other owners in that country as well as in numerous other countries.

Both are broadly based with similar objectives which include:

- meeting and maintaining standards of safety;
- preserving the road network capacity;

- minimising maintenance costs;
- organising preventive maintenance;
- using funding sources efficiently; and
- preserving the investment in structures in whole life terms.

The activities involved in each system are again essentially the same: information gathering, interpretation, prediction, costing and budgeting, planning, optimisation, and decision-making. The major components are somewhat different with different interactions. A general impression can be obtained from Figure 3.4, for DANBRO; this clearly shows the emphasis put on maintaining inventories and how assessment fits into the decision-making process. For major populations of similar types of structure, this breadth is essential.

3.2.2.3 The Swedish Bridge Management System: SAFEBRO

A summary of SAFEBRO is given in reference [3.25]. The overall set up of the system is shown in Figure 3.5, with only five main components:

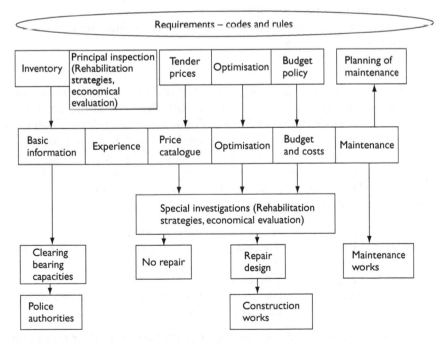

Figure 3.4 Activities included in the DANBRO management system [3.24, 3.25]

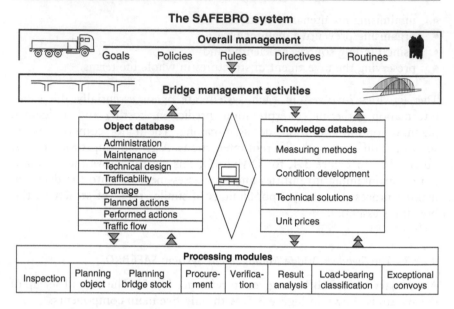

Figure 3.5 The SAFEBRO system [3.25]

(a) Overall management strategy
(b) Bridge management activities
(c) Object databases (the basic inventories for the bridge stock)
(d) Knowledge databases
(e) Processing modules

Strategy/activities is decided at (a) and (b). Processing modules (e) are used to refine and update the core data in (c), while drawing on the knowledge databases (d). The definition and interaction is simple and clear. The system is computer-based, but is supported by guidance documents on particular aspects, for example on inspection regimes [3.26]. This puts great emphasis on the competence and training of personnel, and on the quality and maintenance of documents and records. Five types of inspection, as indicated below with their frequency, are recommended:

Regular Inspection	Routinely, to note unusual damage
Superficial Inspection	Twice a year for major routes, otherwise annually
General Inspection	$\ngtr 3$ years
Major Inspection	$\ngtr 6$ years
Special Inspection	As required to investigate observed defects or critical elements.

Guidance is given on the aims and scope of each type, with clear indications on what to look for.

A common feature of SAFEBRO, and of the other systems described earlier, is the need for sound documentation, typified by the five components shown in Figure 3.5. This must be easily accessible and in a format which permits analysis, evaluation and good decision-making. Having said that, no matter how detailed and elaborate the system may be, there must also be scope for good engineering judgement; in that sense, asset management is no different from design and new construction.

3.2.3 Type 2: Structure specific – other structures

While bridges have received most attention in terms of the systematic development of asset management, approaches have evolved for other types of structure. These may be international, national or owner-driven, often motivated by higher-than-expected maintenance costs (economic), or by concerns over safety, serviceability or loss of function (technical). The purpose of this section is to log a few of these while highlighting some aspects which may be of benefit, if used in a wider context.

3.2.3.1 Multi-storey car parks (MSCP)

In the UK at least, a systematic approach to asset management of multi-storey car parks has gained momentum only relatively recently – possibly triggered by the Pipers Row collapse described in Chapter 2 [2.47]. There are four references which paint the picture:

1. Reference [3.27] This report reviews structural performance and summarises the type of defect found, together with the most significant causes.
2. Reference [3.28] Recommendations for the design and constructions of new car parks, based on feedback from in-service performance.
3. Reference [3.29] Recommendations for the inspection and maintenance of existing car parks.
4. Reference [3.30] A series of short articles, clearly indicating a change of attitude in making car parks more attractive to users, as well as easier to operate and maintain.

In general, multi-storey car parks are open to the elements (Figure 3.6) and therefore subject to the vagaries of wind, rain and temperature. Water is also brought into the car parks by vehicles, sometimes transporting de-icing salts, and there is clear evidence of this from surveys of chloride concentrations beneath vehicle tracks. The general environment is therefore more akin to that of exposed highway and marine structures than that of conventional buildings.

Traditionally, car parks have been designed to the current building Codes, and, therefore, not to realistic exposure conditions. They were often

Figure 3.6 Typical exposed concrete frame car park

designed down to a price, and construction quality was frequently poor, in terms of both concrete quality and achieving the required cover. Ancillary detailing of joints, drainage and waterproofing has also been variable, creating all sorts of water-management problems due to ponding, leaks and run off – in short, ideal conditions for corrosion to occur and for frost action on open ground and top decks. These factors are briefly mentioned here, since they give pointers on what to look for during inspections and assessment; reference [3.28] is an attempt to rectify the situation for new construction.

In terms of assessment, the ICE recommendations [3.29] put the emphasis on the development of Life Care Plans. The essence of these plans is shown in Table 3.5. There are features here common to other systems described earlier, which include:

- a systematic routine approach.
- progressive assessment, as required, from routine surveillance, via condition surveys, to structural appraisal.
- the recognized need for experience and expertise, as the appraisal progresses.

The third point is of special importance for car parks, with their unique characteristics and environmental loadings. Judgement would also be

Table 3.5 Life Care Plan from reference [3.29]

Life Care Plan			
Task	Performed by	Responsible	Programme
Surveillance (Daily)	Car park staff	Car park manager	Daily
Inspection (Routine)	Inspector/Surveyor	Car park manager	Every 6 months
Detailed condition survey	Consulting engineer	Owner/Operator	Every 5 years
Structural appraisal	Consulting engineer	Owner/Operator	Every 10 years
Special inspection	Consulting engineer	Owner/Operator	As required

necessary in terms of timing of the programme, depending, among other things, on:

- structural sensitivity, e.g. the level of redundancy;
- the severity of the local micro-climate;
- the extent and location of defects, with respect to critical sections; and
- the predicted future rate of deterioration.

The recommendations go on to suggest alternative management strategies, with a strong emphasis on preventative maintenance, including better control of water and the local micro-climate. The economic and technical evaluation of alternative repair options is also advocated, with a distinction being drawn between structural repairs and those which are cosmetic or preventative; this is an important point, since survey data [3.27] indicates that many repairs in car parks have either been ineffective or of short duration.

3.2.3.2 Ports, harbours, maritime structures

Quite a lot is known about the durability of maritime structures, in terms of critical local micro-climates, transport mechanisms, and the type of deterioration which can occur (see Chapter 2, e.g. Figure 2.5). However, generally there is less information available on asset management systems.

Exceptions to that are publications from the International Navigation Association [3.31, 3.32]. Reference [3.32] uses Whole Life Costing to link initial design to asset management and relies on bridge experience for much of the technology, for nominal lives in the range 50–80 years. There are, however, differences in outlook and approach mainly because any loss of function, or down time, is borne by the owner through loss of revenue – unlike for bridges, where the users suffer more due to delays. An example is quoted of closing a jetty for 6 days to effect a 20 000 dollar repair, but with a loss of revenue of 1 500 000 dollars. Part of the problem in giving general, detailed guidance is that port infrastructure can vary considerably in character. This is illustrated in simple form in Figure 3.7, taken from reference

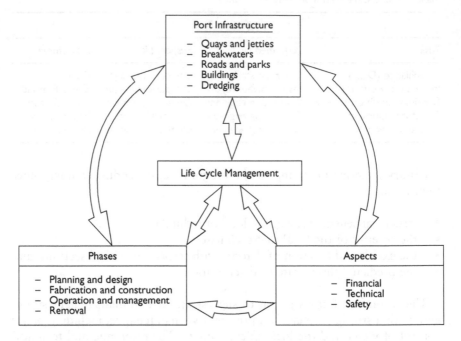

Figure 3.7 Life-cycle management for ports (reference [3.32])

[3.32]. Structured elements can be massive, planar or concrete-framed. Nevertheless, firm guidance is given on the essentials for an asset management system, drawing particular attention to the need for an active computer-based data system, covering both the original structure and any repaired elements. An attractive feature of reference [3.32] is the application of the principles to five major case studies.

3.2.3.3 General civil engineering

Rather like ports and harbours, this is a big group, with a wide diversity of structural types and a range of owners with different needs and strategies. The scope might include major public edifices, utilities, industrial processes, all aspects of water supply and treatment, the retail industry, power supply, sport and leisure, inland transport installations and so on. It is not surprising perhaps that there has been no coherent, focussed development of asset management systems, except in certain sub-groups – and even these often relying on technology transfer from other areas such as bridges.

The group is isolated here for identification purposes, first recognising that it is a major area which should be covered, and that it requires above-average development; second, while the principles from other categories may hold good, some aspects can be very different. This applies to local micro-climates and aggressive actions – most civil engineering structures are exposed; some aggressive actions are unique. As with all assessment work, it is important that the durability 'loads' are clearly identified. Performance requirements are the second major issue, where differences can crop up. Safety will always be a factor, but continuity of function over significant periods may over-ride conventional concerns on serviceability.

An example of a management system for the water power industry can be found in reference [3.25]. The United States Army has also done significant work in this general area, over many years [3.33]. An early example of their attempt to standardise the use of condition rating for decision-making is shown in Figures 3.8 and 3.9. Figure 3.8 shows the basic Condition Index (CI) system, split into three actions zones and using numerical values in the range 0–100. Figure 3.9 is an example of its use, in comparing different maintenance policies.

Zone	Condition index	Condition description	Recommended action
1	85 to 100	**Excellent**: No noticeable defects. Some aging or wear may be visible.	Immediate action is not required.
	70 to 84	**Very Good**: Only minor deterioration or defects are evident.	
2	55 to 69	**Good**: Some deterioration or defects are evident, but function is not significantly affected.	Economic analysis of repair alternatives is recommended to determine appropriate action.
	40 to 54	**Fair**: Moderate deterioration. Function is still adequate.	
3	25 to 39	**Poor**: Serious deterioration in at least some portions of the structure. Function is inadequate.	Detailed evaluation is required to determine the need for repair, rehabilitation, or reconstruction. Safety evaluation is recommended.
	10 to 24	**Very poor**: Extensive deterioration. Barely functional.	
	0 to 9	**Failed**: No longer functions. General failure or complete failure of a major structural component.	

Figure 3.8 Basic Condition Index (CI) system – US Army [3.33]

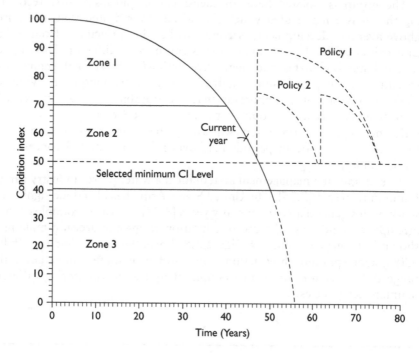

Figure 3.9 Example application for comparing different maintenance policies [3.33]

a. Is it necessary to raise the condition of the facility up to 90 (Policy 1), and likewise, is a CI of 75 high enough (Policy 2)?
b. Is it likely that funds will be available when needed to repeat the rehabilitation in 14 years, as required under Policy 2?
c. What is the likelihood of the CI dropping only five points during the first 14 years under Policy 1?

The system implies monitoring and measurement over significant periods of time, and much will depend on what factors are included in 'condition' and on how the numerical indices are assigned. The approach, with its simplicity and flexibility, and the identification of the three actions zones, has its attractions, particularly in evaluating populations of similar structures. The system is easily capable of further refinement and definition, and variations of it are known to have been used by major owners in North America.

3.2.3.4 Nuclear power installations

This is a sub-group of civil engineering, but is a specialised area, requiring considerable control and rigour within a systematic approach. The full rigour of the system may not always be necessary for less critical civil engineering structures, but development work has been so intense and detailed that useful lessons can be learned. Much of this has happened within the nuclear industry itself, but there has also been considerable activity under the auspices of RILEM. The approach here is not to go into detail, but to give some references where those details may be found.

A key feature in asset management is the discipline imposed by the practice of license renewal. One document setting standards for this is reference [3.34] in association with a computer-based reference manual (COSTAR), which documents current knowledge on degradation and asset management.

A reference document for RILEM activities is given as reference [3.35]. This presents a review of international activities, and has led to a number of workshops between its publication and the present time. It is also understood that RILEM has started new committee activity dealing with the use of concrete in radioactive-waste-disposal facilities.

Two further international players in this field are the OECD [3.36] and IAEA [3.37]. Collectively, references [3.34–3.37] cover current practices, give access to more detailed information and provide a basis for activity in individual countries.

3.2.4 Type 3: Aggressive actions

3.2.4.1 Introduction

The identification and diagnosis of the effects of aggressive actions are dealt with in Chapter 4, and the structural implications in Chapters 5 and 6. However, some existing guidance on aspects of asset management is contained in documents, which are superficially all about individual aggressive actions. These are identified in this section. In assessment work, the priorities relating to aggressive actions are the following:

(a) Being able to identify the action, what causes it, and having some idea of its mechanism;
(b) Being able to diagnose the dominant action, from observed defects and deterioration, which may be due to a number of different factors;
(c) Knowing the nature and scale of the effects that the action can have on the structure and its elements. At a basic level, this reduces to:

  (i) loss of cross-section (concrete or reinforcement);
  (ii) influence on mechanical properties;
  (iii) influence on particular resistances (bending, shear, bond, etc.)
  (iv) the possibility of unforeseen or alternative modes of behaviour due uniquely to the defects.

(d) Knowing what influence the items in (c) above may have on strength, stiffness, stability, serviceability, function, appearance or other performance issues that may be important for different types of structure.
(e) Assessing the time factor. At what rate, may things deteriorate in the future?

The literature is full of papers and reports on all the principal deterioration mechanisms. Much of this is focused on item (a) above, at increasing the understanding and evaluating the relative impact of every conceivable variable in scientific terms. In moving forward from that base towards application in practice, the emphasis has perhaps been more on developing specifications and procedures to minimise the risks for new construction than on the assessment scenario. Nevertheless, much work has been done on item (b), but, in moving through from item (c) to item (e), in engineering terms, there is progressively less universally agreed information available.

However, some such guidance does exist and the sources will be identified for the more important aggressive actions in the sections that follow.

3.2.4.2 Alkali–silica reaction

This mechanism is considered first, not because it is the most important but because it is now well understood, and all of the five items in Section 3.2.4.1 have been covered by authoritative guidance in the UK.

National guidance on the appraisal of existing structures affected by ASR in the UK was first published in 1992 [3.38]. The importance of the expansive nature of the reaction, not only in terms of changes in material properties or cracking, but also structurally in terms of the influence of restraints or the sensitivity of detailing, is known. The basic approach is to derive Structural Element Severity Ratings, as a function of site environment, consequences of failure, expansion index and reinforcement detailing class. However, the report goes beyond that to give guidance on:

(a) how to modify Code equations for design resistance to take account of deterioration due to ASR (especially for bond); and
(b) how to manage ASR-affected structures for different Severity Ratings.

This basic approach has been taken forward, as more evidence became available, and further guidance is given in references [3.15, 3.20]. Particular mention should be made of the CONTECVET Manual [3.39], which modified the approach to operate in terms of restrained, rather than free, expansion and gave more detailed guidance on how Code-based design equations should be modified, including those in Eurocode 2.

3.2.4.3 Corrosion

This is probably the most significant and widespread of all the aggressive actions, and most of the general guidance documents are focused on it, e.g. references [3.10–3.15, 3.20].

Referring to items (a)–(e) in Section 3.2.4.1, (a), (b) and, to a lesser extent, (e) are well covered and understood. There is not the same unanimity of approach to the structural factors (c) and (d). Efforts have been made to develop a full reliability approach (Level 5 in BA 79/98 [3.23] – see Section 3.2.2.1). CONTECVET [3.40] has demonstrated that the Structural Element Severity Rating approach, developed for ASR, can also be used for corrosion, having calibrated it against limited test data. There are now much more experimental data available (e.g. see the References section of [3.20]), and more effort is needed to take this approach forward – structural assessment is still something of a weak link in asset management, when corrosion is involved. Having said that, the relatively simple overall approach in Concrete Society Technical Report 54 [3.14] offers a practical way forward, when supported by an earlier Concrete Society report on cracking [3.41].

3.2.4.4 Freeze–thaw action

Frost action as a deterioration mechanism has a relatively low profile in the UK, but is considered to be significant in certain parts of the world such as Scandinavia, where a great deal of research has been done. It has two principal effects:

(a) Internal mechanical damage; and
(b) Surface scaling, in the presence of saline solutions.

Frost action occurs only when the concrete is effectively saturated. Internal mechanical damage is due to expansive action as ice formation occurs in the concrete, resulting in crack formation, which is quite similar to that due to ASR, or even some forms of sulfate attack. Diagnosis is of great importance here, since the subsequent evolution of damage is quite different for different mechanisms. The practical effect is a loss of compressive and tensile strength plus reductions in E-modulus. The effect on compressive strength can be quite small (say up to 30 per cent), but tensile and bond strengths can reduce by up to 70 per cent, and E-modulus values can be halved.

Initial surface scaling occurs in the cement paste, but can then penetrate deeper into the concrete, particularly if there is also internal mechanical damage, thus dislodging bigger aggregate particles, causing disintegration and reduced cover.

Frost action is mentioned here for two reasons:

1. the importance of diagnosis, to ensure that it is indeed frost action;
2. the fact that structural assessment can be carried out progressively, in applying the same approach and principles as for corrosion and ASR. This is demonstrated in the third of the three CONTECVET manuals [3.42].

3.2.4.5   Different forms of sulfate attack

In the UK, sulfates in soil and groundwater are the chemical agents most likely to attack concrete. Conventional sulfate attack has been recognised for many years and dealt with by defining classes of aggressivity and recommending concrete specifications for each, in Codes and Standards. It has been contended, however, that external sulfate attack is not yet fully understood [3.43].

The situation has become more complicated in recent years, by the discovery of new forms of sulfate attack: delayed ettringite formation [3.44] and thaumasite [3.45], in particular. All of this has led to a re-classification of aggressive chemical environments [3.46].

In assessment terms, diagnosis is again important in determining what type of sulfate attack is occurring, since the physical effects on the concrete may be different for different types of attack. For example, for thaumasite [3.45], it is recommended that all concrete affected by the action should be disregarded in strength and stability calculations, since the concrete can soften into a mush. More generally, the expansive action of sulfate attack can lead to:

– loss of effective concrete cross-sectional area;
– reduced mechanical properties;
– loss of cover to reinforcement;
– loss of bond between concrete and reinforcement;
– possible loss of reinforcement cross-sectional area, due to subsequent corrosion.

3.2.4.6   Summary of 3.2.4 – Aggressive actions

The purpose of Section 3.2.4 is not to give detailed treatment of the different aggressive actions, but to focus on the effects that they may cause in structural performance terms. The key features are given in Section 3.2.4.1. Some indication of how they are covered by current guidance is given in Sections 3.2.4.2–3.2.4.5 for individual aggressive actions.

3.2.5 Type 4 – Inspection and testing procedures

Throughout Sections 3.2.2–3.2.4, the importance of diagnosis has been stressed repeatedly. This relates both to cause and effect and to different facets of the investigation, as assessment progresses. These facets may include:

(1) Identification of the dominant aggressive action
(2) Definition of the environment and local micro-climate (see Table 2.2)
(3) Geometrical and mechanical properties
(4) Nature and scale of defects and deterioration
(5) Structural behaviour, with and without deterioration.

In simple terms, in any assessment, there has to be a balance between Overview and Insight, as illustrated in Figure 3.10.

Overview will always be necessary. How much Insight is required for each of the five facets given above will depend on:

- the level of detail available regarding the asset;
- input from previous surveys/tests/monitoring;
- the nature, scale, and location of the symptoms of deterioration; and
- the type and sensitivity of the structure.

Assessment is seen as being progressive, in moving forward from the preliminary investigation stage only as far as is necessary to take a decision confidently. Moving forward in this way inevitably involves more Insight, to achieve greater accuracy. For facets (1)–(4) above, this generally means more meaningful testing. For facet (5), it may involve more rigorous analysis, or, where deterioration is involved, the interpretation of experimental data from laboratory tests.

The existing general guidance documents provide coverage of test methods, e.g. references [3.12–3.14, 3.22 and 3.25]. Reference [3.22] is particularly attractive, since the scope is set within a defined management system.

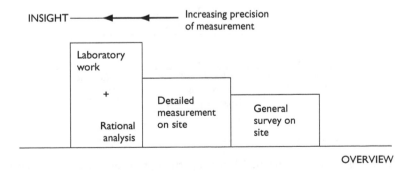

Figure 3.10 The essential balance between Overview and Insight, in assessment

3.3 General principles and objectives

3.3.1 Introduction

The central theme of this book is the management and maintenance of deteriorating concrete structures. The core elements are deterioration mechanism, structural assessment, and repair and remedial options, which follow in Chapters 4–7. Collectively, they could be regarded as an action module – equivalent to the processing modules for the SAFEBRO system in Figure 3.5.

To fully understand the role of the action module, and how it fits into the broader asset management system, it is first necessary to establish some principles and objectives from the range of existing systems illustrated in

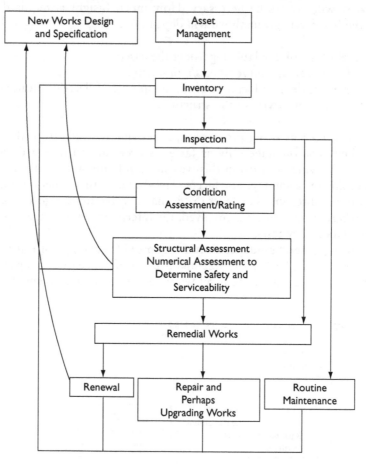

Figure 3.11 Generic action module for inspection, assessment and repair, within an asset management system

Section 3.2. For that the general flow diagrams from *fib* and CEB will be used (Figures 3.1 and 3.2).

3.3.2 Elements/factors in an asset management system

Figure 3.11 is a simplified version of Figure 3.2 and as such represents what we now call the action module. It is also typified by the right-hand side of the broader generic *fib* diagram in Figure 3.1. The question now addressed here is how the remainder of Figure 3.1 might be modified/updated, based on the experience gained from a review of the different management systems discussed in Section 3.2.

It is assumed that the objective of asset management is to maintain an acceptable level of performance, at minimum whole life cost, for the expected life of the structure. Figure 3.1 shows that there are many factors which exert control over that objective, or act as drivers, in terms of how it might be done. Mostly, these are down to the owner, but there are some issues beyond his control, which influence his strategy.

To illustrate this, Table 3.6 has been produced, with two different levels of controlling factors. Level 1 is described as 'national'; as an example, for bridges, there may be over-riding requirements for the road or rail network as a whole, which control what the owner can do. The same holds for the other three issues at level 1.

Level 2 itemises factors and drivers which may influence the owner's strategy more directly; it is a tabular form of Figure 3.1, augmented by some key points taken from the examples given in Section 3.2. The inter-action between those factors is important, and their relative weight may be different for different types of structure, different functions or the size of the population that the owner is dealing with.

In building up such a system, much will depend on what is already there in place. Some basics emerge, however, which include:

a. The need to have a system of graded inspections in place. This cov-ers both the intervals between these inspections and their nature – with guidance on what to look for, and preferably with a standardised approach to reporting findings. There are several examples of this in Section 3.2, most with emphasis on the need for experienced and expert personnel appropriate to each grade (e.g. Table 3.5).

b. The importance of databases/inventories. This is not just record keep-ing, but the creation of an active system, which is updated regularly and which permits interrogation and cross-referencing.

c. The need for support systems or knowledge bases, which is logged in Table 3.6. If the information from (a) and (b) above is to be used effectively, then it requires analysis, as part of decision-making. Some of the examples given will be in-house, but advantage can also be taken of general consensus guidance.

Table 3.6 Factors and drivers which shape an asset management system

	Issues	Factors/drivers
Level 1 (national, federal, local)	National standards/polices Statutory requirements Health and safety Environment/sustainability	
Level 2 (owner)	Future goals and objectives	Useful lives; upgrades; changes in use or expectations (e.g. higher loadings) demolition and replacement.
	Economics	Asset values; budgets; whole life costs; alternative policies/options; insurance and future liabilities, etc.; operational costs.
	Management strategy	Inventories, records, admin; replacement cycles for services, fittings; inspection and maintenance regimes; routine or preventative maintenance.
	Safety	Strength; stiffness; stability; robustness; minimum acceptable performance; assessment methods; reliability; consequences of failure.
	Serviceability/function	Minimise disruption/downtime; inspectability; aesthetics; durability.
	Support systems/knowledge bases	Quality management; improved technology/components; risk management; unit cost; evaluation/analysis/forecasting.

3.3.3 Implications for 'action modules'

Figure 3.11 is a simplistic example of an action module, covering inspection, assessment and repair (if required). To be effective within an asset management system, as outlined in Table 3.6 and backed up by the examples in Section 3.2, some simple principles apply:

1) Clear objectives on what is expected from the action module in terms of:

   - scope and methodology;
   - precision/accuracy, consistent with the chosen method for decision-making.

2) Compatibility with the asset management system as a whole, in terms of policy, organisation and administration and in terms of addressing the key factors and drivers.

3) Economy of use and progressive in approach, in terms of only going as far as is necessary, to permit confident decisions-making.

Application of these three principles imposes a discipline on the action module itself – on its format and approach, and on its interface with the broader management system. To appreciate what this might mean, let us assume the owner wants to know what actions to take in what timescale and what factors he should take into account in reaching a decision. This is illustrated in simple terms in Table 3.7. The factors would include issues other than the results from the assessment, but the latter may well set the timescale.

Table 3.7 Nature and timing of possible actions after assessment

Actions

1.	Do nothing; inspect again in *x* years	
2.	No action now, but monitor	
3.	Routine maintenance; cosmetics; some patch repairs	Evaluate cost/benefits of each in whole life costing terms
4.	Remedial action: specialist repairs and/or protection	
5.	Partially replace, or upgrade, or strengthen	
6.	Demolish and rebuild	

Timescale	**Factors**
Now	Results from assessment
1–5 years	Future change in function
5–10 years	Future change in standards
10–25 years	Type and nature of structure
Longer term	Risk and consequences of failure

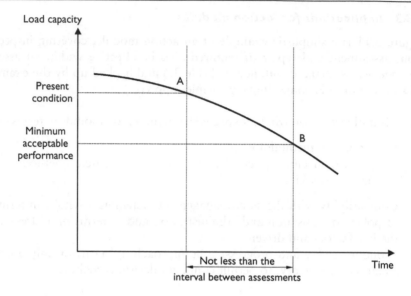

Figure 3.12 Schematic performance profile

If we then assume that the performance profile that emerges from the action module is similar to that shown schematically in Figure 3.12, then several key points emerge. These are outlined below:

(1) 'Performance'
As drawn in Figure 3.12, performance is associated with load capacity, and this will generally be paramount. Even in this limited interpretation, it has to be treated with care, since different forms of resistance (bending, shear etc.) may be affected differently by any given level of deterioration; this possibility is a strong argument for going beyond condition assessment to a more quantified approach.

'Performance' can be extended beyond strength and stability to cover different aspects of serviceability, in fact to any feature where a reduction to a minimum acceptable level would be of special concern to the owner. Examples might include:

 (i) cracking due to corrosion, sufficient to cause delamination or spalling;
 (ii) deformation, which might impair function or operations;
(iii) cracking which is unacceptable for aesthetic reasons or which might lead to leakage;
(iv) a limit on expansion due to ASR, in the presence of restraints, in already highly-stressed regions;
 (v) synergetic effects, e.g. scaling due to frost action leading to an increase in corrosion rate;

(vi) a conservative approach to risk; e.g. a decision to act when surface chloride levels have reached a certain magnitude rather than wait for the critical threshold level and the onset of corrosion.

(2) Minimum acceptable performance levels
Taking decisions on these is crucial in asset management and relates to all the different types of performance identified in (1) above. This affects not only the scope of the work implicit in the action module, but also the nature and precision of the models to be used in evaluating the data, e.g. an ability to predict the onset of cracking due to corrosion or the expansion due to ASR.

(3) The gradient of the performance profile
If performance is assessed only at point A in Figure 3.12, then the action module should have the capability of predicting the gradient between A and B, or at least as far along the time axis until the next scheduled inspection is due.

(4) Progressive assessment
The simple action module in Figure 3.11 depicts a natural progression from inspection through condition rating to structural assessment. The question arises as to how far that path should be followed in individual cases. To answer that, it is necessary to look at Figure 3.11 in a little more detail. This is done in Figure 3.13, derived by the author for CONTECVET [3.38].

Figure 3.11 suggests that there are four levels, where a confident decision might be taken to stop the assessment. Level 1 is obvious. Level 2 suggests a desktop study, to integrate as-built records (if available) into a re-evaluation of its design basis, while looking at structural sensitivity. Stopping at this level is likely to happen only if conservative analyses indicate a considerable margin of safety, and the rate of deterioration is low, in relation to the inspection intervals.

Level 2 is more likely to be a preliminary to Level 3, perceived as a formal condition rating. There are several examples of such ratings in Section 3.2, and this aspect will be dealt with later in Chapters 5 and 6, but it should be noted that calculations are recommended even at this early stage, as an aid to decision-making.

'Detailed Investigation' in Figure 3.13 is equivalent to 'Structural Assessment' in Figure 3.11, i.e. analysis/calculations targeted at specific aspects of safety and serviceability (Level 4). Note the important point made in the figure, of the need to establish minimum performance requirements before undertaking a Level 4 investigation.

(5) Detailed structural investigation
Current practice reveals a wide range of approaches to the numerical structural assessment of deteriorating concrete structures. Considerable research effort has gone into the development of full reliability analysis [3.47, 3.48].

Other approaches have involved modifications to design equations (and to their inputs), on the assumption that the level of reliability from these semi-probabilistic methods has been generally accepted; references [3.9, 3.11, 3.25 and 3.38–3.40] are all in this category.

Possibly the most practical way forward is the progressive approach developed for UK highway bridges [3.15, 3.23], in moving from simple analysis towards full reliability analysis, should the latter prove necessary. This will be developed in more general terms in Chapters 5 and 6.

(6) Management aspects

Although the procedure in Figure 3.13 seems relatively straightforward, situations can arise where decision-making is not clear-cut. This is illustrated schematically in Table 3.8. The situation shown for Preliminary Condition

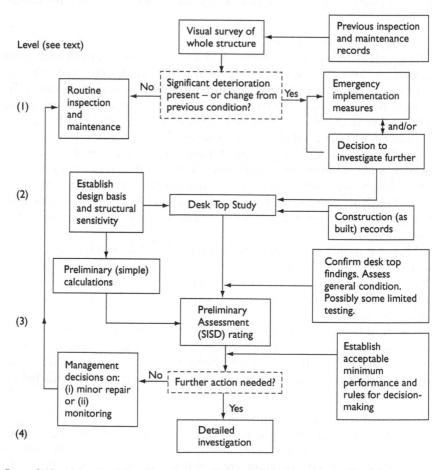

Figure 3.13 Action module – flow diagram for progressive assessment [3.39]

Table 3.8 Management aspects arising from progressive assessment (Figure 3.13)

Assessment level	Conclusion Based on	Result	Reason	Recommendations
3. Preliminary condition rating	Records Inspection data and measurements	Adequate	Sufficient residual life and load-carrying capacity	Monitor. Review Inspection Regimes.
	Nature, location and extent of deterioration	Borderline	Insufficient data, or estimated performance marginally below that required.	Move on to a Detailed Assessment. Further testing.
	Simple analyses Condition rating Structural sensitivity	Inadequate	Insufficient residual life and load-carrying capacity.	Possibly modify adequacy criteria and re-assess. Immediate Detailed Assessment. Consider alternative remedial measures – and their urgency.
4. Detailed structural assessment	As Preliminary, plus: Monitoring	Adequate	Sufficient capacity for required loading.	Monitor. Revise Inspection Regimes.
	Laboratory tests More rigorous analyses (i.e. more INSIGHT, see Figure 3.11)	Borderline	Insufficient data, or residual capacity marginally low, or projected high future deterioration rates.	Possibly modify adequacy criteria and/or evaluate actual loading, and re-assess. Consider alternative remedial measures. Additional testing – including possible load test.
	Deterioration rate Acceptable minimum performance	Inadequate	Insufficient capacity on one or more aspects of performance.	Re-assess, as per borderline situation. Take decision on nature and timing of optimum remedial action. Short-term action needed? – e.g. propping; load restrictions, etc.

Rating is a simplification of that in Figures 3.8 and Figure 3.9, and for a result other than 'Adequate', detailed assessment is inevitable.

Detailed assessment, Level 4, is different in the sense that a decision/action has to be taken, unless the minimum performance criteria can be reviewed and relaxed, and the change justified. Table 3.8 is introduced at this time to make the point that assessment is not an isolated entity, but a tool to be used as part of the overall asset management system.

References

3.1 O'Brien T.P. Durability in design. *The Structural Engineer*. Vol. 45. No. 10. pp. 351–363. IStructE, London, UK. October 1967.

3.2 British Standards Institution (BSI). British Standard Code of Practice CP3 – Chapter IX (1950). Code of functional requirements of buildings. Chapter IX: Durability. BSI, London. 1950.

3.3 The Building Centre. *Maintenance manual and job diary*. The Building Centre, London, UK. 1966.

3.4 NBA Construction Consultants. *Maintenance cycles and life expectancies of building components and materials; a guide to data and sources*. NBA, London. June 1985.

3.5 British Standards Institution (BSI). BS *Guide to building maintenance management. BS 8210: 1986*. BSI, London, UK.

3.6 International Organisation for Standardisation (ISO). *Bases for design of structures – assessment of existing structures ISO 13822: 2001 (E)*. Contact via BSI.

3.7 British Standards Institution (BSI). *Durability of buildings and building elements, products and components. BS 7543: 1992*. BSI, London, UK.

3.8 International Organisation for Standardisation (ISO) *Buildings and constructed assets – service life planning – Part 1: General principles. BS ISO 15686–1: 2000*. BSI, London, UK.

3.9 Institution of Structural Engineers. *Appraisal of existing structures*. 1st Edition 1980. 2nd Edition 1996. IStructE, London, UK.

3.10 American Concrete Institute (ACI) *Strength evaluation of existing concrete buildings*. ACI 437R-91 (reapproved 1997). ACI, Detroit, USA.

3.11 Canadian Standards Association ASSOCIATION (CSA). Existing Bridge Evaluation to CSA Standard CAN/CSA-S6-88, Design of Highway Bridges. Supplement No 1, CSA, Rexdale, Toronto, Canada. 1990.

3.12 Comite Euro-Internationale Du Beton (CEB). *Strategies for testing and assessment of concrete structures*. Bulletin 243, CEB, Lausanne, Switzerland. May 1998.

3.13 Federation Internationale Du Beton *(fib)*. *Management maintenance and strengthening of concrete structures. Fib* Bulletin 17. *fib*, Lausanne, Switzerland. April 2002.

3.14 The Concrete Society. *Diagnosis of deterioration in concrete structures: identification of defects, evaluation and development of remedial action*. Technical Report 54. 2000. p. 72. The Concrete Society, Camberley, UK.

3.15 The Highways Agency. *Design Manual for Roads and Bridges (DMRB)*. Website. http://www.official-documents.co.uk/documents/deps/ha/dmrb/index.htm.

3.16 Flint A.R. and Das P. Whole life performance-based assessment rules – background and principles. *Proceedings of a conference on safety of bridges*. July 1996. Thomas Telford Ltd, London, UK.

3.17 Parsons Brinkerhoff. *A review of bridge assessment failures on the motorway and trunk road network*. Final report. Contract 2/419. The Highways Agency, London, UK. December 2003.

3.18 County Surveyors Society (CSS). *Funding for bridge maintenance*. CSS Bridges Group, London, UK. February 2000.

3.19 Concrete Bridge Development Group (CBDG). *The assessment of concrete bridges (1)*. Report of a Task Group. CBDG, Camberley, UK. 1997.

3.20 Concrete Bridge Development Group (CBDG). *Notes for guidance on the assessment of concrete bridges*. Publication No. 9. CBDG, Camberley, UK. 2005.

3.21 W.S. Atkins et al. *Management of highway structures: a code of practice*. ISCN 0115526420. The Stationery Office, UK. September 2005.

3.22 Concrete Bridge Development Group (CBDG). *Guide to testing and monitoring the durability of concrete structures*. Technical Guide 2. CBDG, Camberley, UK. 2002.

3.23 The Highways Agency. *The management of sub-standard highway structures*. BA79/98. 1998 (incorporating Amendment No 1, dated August 2001). Highways Agency, London, UK. (Also, see [3.15].)

3.24 BRIME (Bridge Management in Europe). *Deliverable D14*. Final Report. March 2001. www.trl.co.uk/brime/deliver.htm.

3.25 REHABCON. Project No. IPS – 2000 – 0063; Innovation and SME Programme. Strategy for maintenance and rehabilitation in concrete structures. Guidance Manual. www.rehabcon.org/home/login.asp.

3.26 Swedish National Road Administration (SNRA). *Bridge Inspection Manual Publication 1996 – 036 (E)*. SNRA, Borlange, Sweden. June, 1996.

3.27 Henderson N.A., Johnson R.A. and Wood J.G.M. *Enhancing the whole life structural performance of multi-storey car parts*. Report to the Office of the Deputy Prime Minister, London. September 2002.

3.28 Institution of Structural Engineers. *Design of underground and multi-storey car parks*. 3rd Edition. IStructE, London, UK. 2002.

3.29 Institution of Civil Engineers. *Recommendations for the inspection and maintenance of car park structures*. ICE, London, UK. 2002.

3.30 The Concrete Society. Feature on car parks. *Concrete*. Vol. 30. No 4. pp. 35–49. The Concrete Society, Camberley, UK. April 2005.

3.31 International Navigation Association. *Inspection, maintenance and repair of maritime structures, exposed to material degradation caused by salt water environments*. PIANC PTC 11: Working Group 17 report. PIANC General Secretariat, Brussels, Belgium. 1990.

3.32 International Navigation Association. *Life cycle management of port structures: general principles*. PIANC PTC 11: Working Group 31 report. PIANC General Secretariat, Brussels, Belgium. 1998.

3.33 US Army Corps of Engineers. *Evaluation and repair of concrete structures*. Report EM 1110-2-2002. Washington DC, USA. June 1995.

3.34 Electrical Power Research Industry (EPRI). *PWR Containment Structures License Renewal Report*. Ref. No. EPRI TR-103842. EPR, Palo Alto, California, USA. July 1994.

3.35 Naus D. (Editor) *Considerations for use in managing the aging of nuclear power plant concrete structures*. RILEM Technical Report No 19 [NEA/CSNI/R (95) 19]. RILEM, Issy-les-Moulineaux, France. November 1995. Website: www.nea.fr/html/msd/docs.

3.36 Organisation of Economic Cooperation and Development (OECD). *Report of Task Group reviewing activities in the area of aging of concrete structures used to construct nuclear power plant fuel cycle facilities*. Report NEA/CSNI/R (2002) 14. Nuclear Energy Agency, OECD, Paris, France. July 2002.

3.37 International Atomic Energy Agency (IAEA). *Assessment and management of aging of major nuclear power plant components important to safety*. IAEA – TECDOC – 1025. IAEA, Vienna, Austria. June 1998.

3.38 Institution of Structural Engineers. *Structural effects of alkali–silica reaction: technical guidance on the appraisal of existing structures*. IStructE, London, UK. July 1992.

3.39 CONTECVET (BRITE-EURAM PROJECT IN30902D). *A validated user manual for assessing concrete structures affected by ASR*. Available from the British Cement Association, Camberley, UK. 2000.

3.40 CONTECVET (BRITE-EURAM PROJECT IN30902D). *A validated user manual for assessing corrosion-affected concrete structures*. Available from the British Cement Association, Camberley, UK. 2000.

3.41 The Concrete Society. *The relevance of cracking in concrete to corrosion of reinforcement*. Technical Report 44. 1995. The Concrete Society, Camberley, UK.

3.42 CONTECVET (BRITE-EURAM PROJECT IN30902D). *A validated user-manual for assessing concrete structures affected by frost*. Available from the British Cement Association, Camberley, UK. 2001.

3.43 Neville A. *The confused world of sulfate attack on concrete*. Cement and Concrete Research. 34(2004). pp. 1275–1296. Elsevier Ltd.

3.44 Building Research Establishment. Delayed ettringite formation: in situ concrete. *Information Paper IP 11/01*. BRE, Garston, UK. 2001.

3.45 Department of Environment, Transport and the Regions (DETR). *The thaumasite form of sulfate attack: risks, diagnosis, remedial works and guidance on new construction*. Report of the Thaumasite Expert Group. DETR, London. January, 1999.

3.46 Building Research Establishment. Concrete in aggressive ground (4 parts). *BRE Special Digest 1 (2nd Edition)*. BRE, Garston, UK. 2005.

3.47 BRITE-EURAM PROJECT BE 95-1347. *Duracrete – probabilistic performance – based durability design of concrete structures*. Website: www.bouwweb.nl/cur/duracrete.

3.48 BRITE-EURAM PROJECT BET2-0605. Duranet – network for supporting the development and application of performance based durability design and assessment of concrete structures. Website: www.duranetwork.com.

Chapter 4

Defects, deterioration mechanisms and diagnosis

4.1 Introduction

This chapter is concerned with the factors which can contribute, singly or in combination, to the deterioration of concrete structures. A major concern is their identification from inspections and diagnostic testing. In terms of the action module defined in Chapter 3, it covers the procedures associated with level 1 in Figure 3.13 (repeated here for convenience as Figure 4.1), while moving towards level 2, in preparation for a formal condition rating (level 3).

Figure 4.1 shows a dependence on previous records and a starting point represented by a visual survey. This will entail the careful logging of all visible defects and signs of deterioration, with the emphasis on the nature, extent and location of these. For many structures, this is not an easy task, for a number of reasons:

(i) difficulties regarding access, especially to hidden zones;
(ii) deciding what constitutes a defect, and classifying these in terms of type, magnitude, location, possible significance and likely cause;
(iii) settling on the extent and scope of the study, and on whether or not to investigate further, via some form of site or laboratory testing.

In moving progressively forward in this way, it is important to have clear objectives, while remembering that it may be feasible to take decisions at levels 1, 2 or 3 in Figure 4.1. If the decision is taken to go as far as a formal condition rating at level 3, then factors additional to the simple recording of defects come into consideration. These include:

(a) logging and classifying the environment and local micro-climates, since this could be important once the primary deterioration mechanism has been identified;
(b) getting to know more about the structure itself, as built, e.g. actual cover values and some indication of concrete quality;

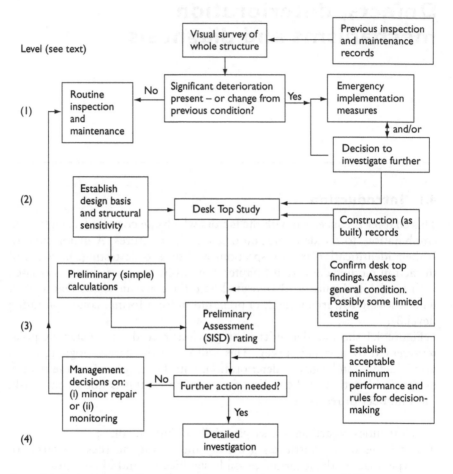

Figure 4.1 Action module flow diagram for progressives assessment [4.9]

(c) studying the condition and functioning of features such as joints, bearings, drainage, waterproofing/protective systems, etc.

These factors are important in themselves, and could become more important later, when considering options for remedial action. Also, even at this early stage in the action module, it is worth remembering the five priorities, (a)–(e), identified in Section 3.2.4.1, should it prove necessary to take the assessment as far as a full structural investigation (level 4 in Figure 4.1).

In the preliminary stages of the action module, the focus will generally be on identifying the dominant deterioration mechanism from the observed effects. While the nature of these effects for suspected mechanisms may generally be known from the literature, they may be distorted or exacerbated

Table 4.1 Some examples of defects and actions which may affect a primary deterioration mechanism

Category 1 *Actions or defects affecting the concrete*	Category 2 *Non-structural and structural cracking*	
Weathering	Plastic settlement	
Abrasion	Plastic shrinkage	
Leaching	Early age thermal effects	
Honeycombing	Long-term shrinkage	
Pop-outs	Creep	
	Ambient temperature – movement and restraint	
	– internal temperature gradients	
	Design loads	
	Settlement	
	Restraints	– determinacy
		– non-structural elements

by defects due to other causes. Diagnosis therefore requires both expertise and experience. Some simple examples of this are given in Table 4.1, in two separate categories. Those in Category 1 tend to damage or reduce the outer surface of the concrete, either physically or in terms of concrete quality. Leaching can also increase the rate of carbonation or chloride penetration.

Most concrete structures are subject to cracking at some stage in their lives. The examples given in Category 2 may:

- occur in different timescales;
- be permanent or transitory; and
- be dormant or live.

An awareness of the characteristics of this type of defect, and how it might influence or distort the effects of a primary deterioration mechanism, is therefore important in preliminary investigations.

The key objective at this stage is the unambiguous identification of the dominant aggressive action, since an option in remedial terms may be to stop or slow down any further deterioration. It may also be necessary to consider future synergetic effects where more than one mechanism is significant. Some possible examples of synergy are given in Table 4.2. Clearly identifying the underlying causes of observed effects is important in predicting the nature of likely future damage and hence in selecting the most appropriate remedial measure.

At face value, Figure 4.1 looks relatively straightforward in moving towards levels 2 and 3. In practice, it is not. Words and phrases like 'significant', 'decision to investigate further', 'assess general condition',

Table 4.2 Some examples of synergy due to deterioration mechanisms acting simultaneously

Combination of mechanisms	Possible effects
Surface scaling due to frost and corrosion	This may lead to a gradual reduction of the cover to the reinforcement and, hence, increases the likelihood of corrosion.
Alkali–silica reaction, and either frost action or corrosion.	The expansive action of ASR may lead to wide cracks which can fill with water, and which, if frozen, may cause internal mechanical damage. This same action may also permit easier access to the reinforcement for water containing chlorides, causing more severe corrosion. On the other hand, gel caused by ASR may fill pores, thus densifying the cement matrix.
Leaching and frost reaction	The influx of water may increase the moisture uptake and, hence, reduce the internal frost resistance.
Leaching and corrosion	The leaching of lime from the concrete cover increases the rate of carbonation and the diffusivity of chlorides and reduces the critical threshold level.

'possibly some limited testing' have to be addressed and decisions made. There are no general hard-and-fast rules here, which can be applied universally. Much will depend on the type of structure and its environment, the information available from all sources at level 1, and which primary deterioration mechanism is suspected.

Basically, this is progressive problem-solving, gathering information from observations and relevant tests – and evaluating that with judgement and a sense of perspective. There are very many measurements that can be made and tests that can be used. In pursuing defined objectives, judgement comes in deciding which ones to use, and perspective, in being aware of the accuracy, variability, reliability and relevance of the results obtained.

In the author's experience, good practical guidance is given in references [4.1] and [4.2]. One deterioration mechanism, where procedures are well-established is ASR. Figure 4.2, taken from reference [4.3], gives an overview of the recommended diagnostic process; the clause numbers shown are those in reference [4.3]. Figure 4.2 is included to provide a template of the thought process involved, starting from an early presumption that the observations made on site were indicative of ASR. It is not a universal panacea for all possible aggressive actions, especially at a detailed level. Much will also depend on whether or not the investigation is a one-off or whether monitoring is possible, with observations made at intervals, preferably in varying ambient conditions.

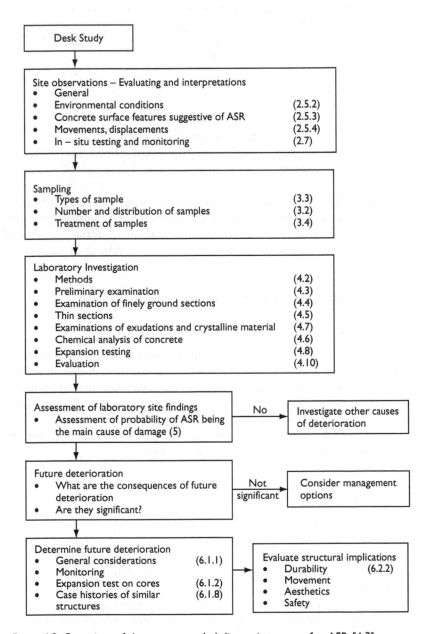

Figure 4.2 Overview of the recommended diagnosis process for ASR [4.3]

Note: Section numbers refer to BCA report: Diagnosis of alkali–silica reacion (4.3).

Against this cautionary introduction, the purpose of this chapter is to establish some basic principles, in progressively following Figure 4.1. The associated guidance will inevitably be general, and where full treatments are not provided, e.g. on test methods, references will be provided for the detail. The sequence in doing this will be:

4.2 Identification of defects and their causes
4.3 Aggressive actions
4.4 Influence of the environment and micro-climate
4.5 Inspection, testing and preliminary diagnosis
4.6 Interpretation and evaluation; pre-preparation for preliminary assessment References

The plan is that this chapter should cover the issues and processes involved in moving towards the desktop study in Figure 4.1, and in preparation for the level 3 assessment, which is dealt with in Chapter 5, followed, if necessary, by the level 4 detailed assessment – the subject of Chapter 6.

4.2   Identification of defects and their causes

4.2.1   Introduction

In undertaking a visual inspection, engineers may be confronted by deficiencies arising from a number of causes, with deterioration due to an aggressive action being intermingled with defects which are construction-related. The different causes have to be identified, and the purpose of this section is to give guidance on how that might be done. For practical reasons, therefore, the scope covers all visible deficiencies, but with more detail on the effects of aggressive actions appearing later in Section 4.3.

4.2.2   Cracking

The tensile strength of concrete is low compared to its compressive strength. This is recognised in design, when providing resistance to imposed loads, by limiting the width of structural cracks. Provided there is adequate cover to the reinforcement and the concrete is properly specified and placed, this is regarded as being acceptable and of no major threat to reinforcement corrosion – except possibly in severe exposure conditions in the presence of chlorides.

Cracking in reinforced concrete can also originate from numerous other sources. These are generally referred to as intrinsic or non-structural, and a common classification for these is shown in Figure 4.3 [4.4, 4.5]. In general, the causes are attributed to chemical or physical changes taking place within the concrete, such as restraint of plastic shrinkage, thermal movements and

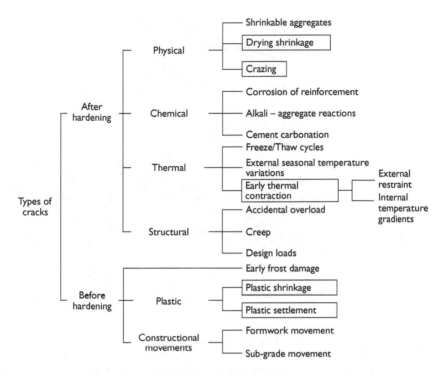

Figure 4.3 Types of crack (Reprinted by permission of The Concrete Society. All rights reserved [4.4, 4.5])

expansive processes. Strain or movement is central to most forms of cracking, and its magnitude can depend on the action of external or internal restraints. Some aggressive actions have been inserted into this general classification (corrosion, ASR, freeze–thaw). Cracking due to corrosion occurs parallel to the reinforcement as the expansive corrosion products cause the concrete cover to crack or even spall. Cracking due to the other two aggressive actions is also essentially due to expansion; each has its own basic characteristics, which may be modified by the presence of reinforcement.

Both references [4.4] and [4.5] contain a diagram of a hypothetical structure, showing where the different types of crack are most likely to occur; this is reproduced here as Figure 4.4.

In this chapter, it is not proposed to give a detailed treatment of the formation mechanics of the different types of crack; that detailed information can be found in references [4.1, 4.4] and [4.5]. Here, in assessment, the key point is one of recognition, since it is important to distinguish between cracks which have occurred due to an earlier event, and those which relate to an ongoing problem such as corrosion. Some cracks occur very early in

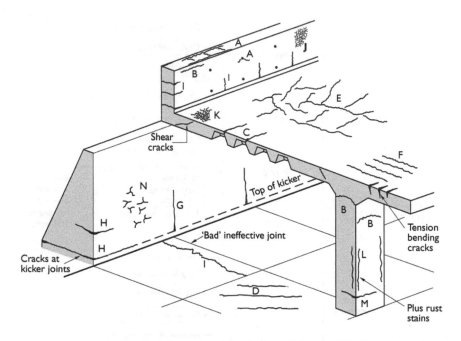

Figure 4.4 Examples of intrinsic cracks in a hypothetical structure [4.4]. Type of cracking: A, B, C = Plastic settlement; D, E, F, = Plastic shrinkage; G, H = Early thermal contraction; I = Long term drying shrinkage; J, K = Crazing; L, M = Corrosion of reinforcement; N = Alkali-silica reaction

the life of a structure (e.g. plastic settlement and shrinkage cracks, and early thermal contraction) and do not develop further once they are formed. As such they do not pose a direct threat to the integrity of the structure, but may cause future problems as potential corrosion sites.

4.2.3 Other sources of defects

A selection of these is presented in Table 4.3, under two separate headings: construction-related and in-service related.

The list is indicative only. Some construction-related defects may be cosmetic only, but with the potential to be sites where effects due to aggressive actions are initiated. The assumed durability characteristics of the concrete may also be impaired, e.g. due to poor compaction or loss of grout. Potentially the most serious defect is low cover in a corrosive environment – something that should always be looked at if there are signs of corrosion (rust stains, cracking, etc.).

The nature/type of in-service defect depends on the environment and the function of the structure. Salt weathering, leaching and abrasion can all occur in marine or maritime structures, for example. Surface scaling may

Table 4.3 Some sources of defects in concrete structures

Construction-related	In-service related
Low cover	Weathering
Variable compaction/curing	Abrasion
Blow holes/voids	Erosion/cavitation
Honeycombing	Impact damage
Excessive bleeding	Joint leakage
Grout loss	Surface scaling
Construction joints	Leaching
Exposed spacers/tying wire	Exudations/surface defects

occur due to frost action; erosion or cavitation, due to the action of flowing water. Leaking joints can be a problem in many types of structure, since this affects both moisture content and the form that water can take on the surface of the structure, e.g. run-off or ponding. The importance of moisture state, diurnal and seasonal, cannot be over-estimated when considering the effects of aggressive actions, and the defects recorded here can have a major influence on that.

4.3 Aggressive actions

4.3.1 Introduction

Like most of the common construction materials, concrete is prone to either physical or chemical attack. The sources of the aggressive actions are generally external, but concrete is perhaps unique in that some significant actions may be initiated by the characteristics of the mix constituents themselves. If this latter situation is suspected, it is prudent to consider the age of the structure and exactly what has gone into its construction, commensurate with practice at the time, since the constituents may be variable and concrete technology has changed over the years. A detailed description of all possible aggressive actions is beyond the scope of this section, and indeed of this book. The purpose is to provide a perspective, through a process of identification and association with different types of structure, prior to focusing on the more significant and widespread actions. While some information will be given on these, heavy emphasis is put on the use of references, for readers wishing for greater detail and understanding. What follows should be read in conjunction with Section 3.2.4.

4.3.2 Overall perspective

Table 4.4 is an extension of the one due to Buenfeld [4.6]. It gives an indication of where different aggressive actions are most likely to occur.

Table 4.4 Overview of aggressive actions linked to type of structure

Legend:
- ■ (dark shading) Indicates likely dominant deterioration mechanism
- ▥ (vertical lines) Indicates mechanism can occur and may be significant

	Cl – induced corrosion	CO₂ – induced corrosion	Sulfate attack	Frost action	Abrasion	Leaching	ASR	Acids	Others
'Normal' above-ground buildings		■					▥		
Multi-storey car parks	■				■		▥		
Foundations			■				▥		
Bridges and road-side structures	■	▥		▥			▥		
Marine and maritime structures	■			▥			▥		
Dams				▥		▥	■		
Agricultural structures		■					▥	■	■
Industrial processes		▥			■			■	▥
Tunnels	■		▥						▥
Tanks and pipes	▥					■		■	▥

The cross-hatched zones indicate where a particular action is likely to be the most significant and controlling deterioration mechanism, given sufficient intensity and the right micro-climate. The zones with horizontal hatching are those where the action can occur and possibly be significant but less likely to be dominant. Common sense is required in reading this table. If chlorides, in whatever form, are not present, then chloride-induced corrosion will not occur; if the ambient temperature rarely falls below zero, and the concrete is not saturated, then frost action will not occur; what happens with tanks and pipes depends on what they are carrying and so on. The table is only a guide.

While examining Table 4.4, it is worth reflecting on the type of damage/deterioration that can occur. Abrasion is perhaps unique, in being a physical effect due to external forces; the concrete surface gets worn away, as might occur in car parks, factories, or in maritime structures due to wave action involving sand and pebbles. Forms of abrasion can also occur in structures such as spillways and in tunnels and pipes having a high flow rate; this can lead to erosion and cavitation.

Frost action can also be isolated in the table. This occurs when water held in the capillary pores in the concrete freezes at low temperatures. If the concrete is saturated, the resulting expansive forces can lead to internal micro-cracking (thus reducing the concrete's mechanical properties) and surface scaling, or even disintegration to the depth to which freezing conditions have penetrated. Repeated freeze–thaw cycles can cause cumulative damage. The surface scaling can get exacerbated if the concrete freezes while in contact with saline solutions, thus accelerating the reduction of concrete cover.

For the remaining aggressive actions in Table 4.4, there is common ground, namely the transport of ions, gas or water. The presence of water is crucial for most chemical reactions to be significant. Chemical attack on the concrete generally occurs via micro-structural change and the decomposition of the products of hydration; this involves reactions between the constituents of the concrete and the penetrating action. Corrosion due to carbonation of the concrete also falls broadly in this category in reducing the protective alkalinity; even chloride-induced corrosion may be so classified, should the sources of the chlorides be internal either from the aggregates or from the use of an accelerator such as calcium chloride. More commonly, of course, the sources of chlorides are external, 'augmenting' the importance of transport mechanisms through the concrete to the reinforcement.

The general point being made above is that the constituents of the concrete mix are important in most forms of chemical attack and will have to be investigated during the diagnostic stage, partly to be sure of the nature of the degradation process and partly to study the likely future rate of deterioration. It should also be recognised that all deterioration mechanisms involve several different interacting processes, and some understanding of this is important, not only for diagnosis but also for assessment, affecting decision-making on the nature and timing of remedial action.

For the above reasons, before briefly covering some of the more important aggressive actions, it is necessary to provide a perspective of the role of material characteristics, particularly of cements and also of aggregates. This follows in Section 4.3.3.

4.3.3   Material considerations

This section does not purport to be a treatise on the properties of cement, concrete, reinforcement and prestressing steel. The intention is to provide a perspective with emphasis on the characteristics of the mix ingredients which play a part either by being integral to particular forms of deterioration or by their ability to resist aggressive actions from external sources. The section is subdivided as follows:

- cements
- aggregates
- concrete
- reinforcement and prestressing steel

In each case, references are given for those interested in following up in detail.

4.3.3.1   Cements

The basic ingredients of concrete are fine and coarse aggregate, plus cement and water which take part in a chemical hydration process whose products provide a firm and hard mass, via the hydrated cement paste binding the whole mix together.

The cement may be Portland cement or a mixture of Portland and other hydraulic cement or pozzalamic material, the most common of which are blastfurnace slag, fly ash or silica fume. The main constituents of Portland cement are lime, silica, alumina and iron oxide, with other minor compounds, the most interesting being the oxides of sodium and potassium (the alkalis).

The composition of cement has changed a great deal over the years, a fact to remember when investigating the present and future condition of old concrete structures. There are several reasons for this, the more important ones being:

(a) improvement and changes in the manufacturing process itself;
(b) a demand from users for stronger cements;

(c) a demand for higher early strengths, say in a precast factory, or to permit the earlier striking of formwork on-site;

(d) special requirements for particular forms of construction, e.g. reduced heat of hydration in massive sections, or easier placing conditions when pumping is to be used or where there is heavy congestion of reinforcement;

(e) requirement for improved resistance to specific forms of aggressive actions, e.g. sulfates, or cements having lower alkali contents.

The net effect of this has been stronger cements, more finely ground, with a range of chemical compositions and alkali levels and with varying heat of hydration. 'Brand' names abounded, including:

− Ordinary Portland cement (OPC)
− Rapid hardening Portland cement (RHPC)
− Low heat Portland cement (LHPC)
− Sulfate resistant Portland cement (SRPC)
− Portland blastfurnace cement (PBC)

This was a process of evolution, based on supply and demand, involving the pursuit of greater economy and better technical performance in a supply chain which has changed significantly in the UK since the 1950s. Reviews of the significance of these changes did not really take place until the 1980s, when concerns on durability were on the increase. The Concrete Society provided such a review in 1987 [4.7]. This confirmed the general increase in strength and a rise in the ratio of 7- to 28-day strengths. It was also possible to achieve a given workability and concrete strength with lower cement contents and higher water/cement ratios – conceivably producing more permeable concretes.

This evolutionary process has led to the current (2005) situation shown in Table 4.5. This is derived from BS 8500 [4.8], the complementary British Standard to BS EN 206-1, the European Standard for concrete.

It may be seen that evolution in the UK has led to the increased use of cements containing either ground granulated blastfurnace slag or fly ash. BS 8500 then goes on to give detailed guidance on type of cements that should be used for defined exposure classes. This consolidation of many years' research and practice should mean improved technical performance in durability terms for current and future structures. In assessment, however, there is still the need for awareness of the characteristics of the cements in use at the time of construction.

For detailed references on cements, the author has relied mostly on the book by Neville [4.9]. This comprehensive volume covers not only the basics of cement and concrete, but also has chapters on different aspects of

Table 4.5 BS 8500 [4.8] Cement combinations and designations (Permission to reproduce extracts from the British Standard is granted by BSI)

Designation	Composition	Cement/combination Type (BS 8500)
CEM I	Portland cement	CEM I
SRPC	Sulfate-resisting Portland cement	SRPC
IIA	Portland cement with 6–20% of fly ash (pfa), ground granulated blastfurnace slag or limestone[a]	CEM II/A-L CEM II/A-LL CIIA-L CIIA-LL CEM II/A-S CIIA-S CEM II/A-V CIIA-V
IIB	Portland cement with 21–35% of fly ash (pfa) or ground granulated blastfurnace slag	CEM II/B-S CIIB-S CEM II/B-V CIIB-V
IIB+SR	Portland cement with 25–35% of fly ash (pfa) or ground granulated blastfurnace slag	CEM II/B-V+SR CIIB-V+SR
IIIA	Portland cement with 36–65% ground granulated blastfurnace slag	CEM III/A[b] CIIA[b]
IIIB	Portland cement with 66–80% ground granulated blastfurnace slag	CEM III/B CEM III/B L CIIIB
IIIB+SR	Portland cement with 66–80% ground granulated blastfurnace slag where, if the alumina content of the slag exceeds 14%, the C3A content of the Portland cement fraction does not exceed 10%	CEM III/B+SR[b] CIIB+SRB[b]
IVB	Portland cement with 36–55% of fly ash (pfa)	CEM IV/B PIV/B-V CIVB
IVB+SR	Portland cement with 36–40% of fly ash (pfa)	CEM IV/B PIV/B-V+SR CIVB

Notes
a There are a number of other second main constituents, but due to costs these will only be used when specifically specified, i.e. silica fume and metakaolin.
b Inclusive of low early strength option.

durability. Further, it is now in its fourth edition, having been first published in 1963, and therefore provides a record of how cements have changed since that time. Possibly, more related to current types of cement, and to the influence of these on corrosion resistance in particular, is Concrete Society Technical Report 61, written by P. B. Bamforth [4.10].

The emphasis in this book is on the cements listed in Table 4.5, or their predecessors (the so-called 'brand' names), since these represent the bulk of concrete practice. It should be noted, however, that other special cements have been developed over the years for specific purposes. These include ultra high early strength cement, masonry cement, white cement and even the highly specialised oil-well cement. Two of these are singled out here, since they have been used in the past for special structural applications.

The first is supersulfated cement (SSC), manufactured from granulated blastfurnace slag (80–85 per cent), calcium sulfate (10–15 per cent) and Portland cement clinker (up to 5 per cent). This was resistant to attack in aggressive conditions and was certainly used in the manufacture of concrete pipes, placed in contaminated, acidic or sulfate-rich ground. It may also have been used in railway overbridges built in the era of the steam train. SSC has a low alkali content, providing reduced resistance to corrosion of embedded metal. It also suffered a loss of strength when carbonation occurred. Its low heat of hydration may also mean that it could have been used in large concrete pours. SSC is no longer manufactured in the UK.

Next, mention must be made of high-alumina cement (HAC). Its principal components are alumina and lime, with about 15 per cent of ferrous and ferric oxides, around 5 per cent of silica, and traces of titanium oxides, magnesia and alkalis. Its origins were in France in the early 20th century, and its sulfate-resisting properties led to its being used in bridges in the UK in the first half of that century. Its main advantage is a very high rate of strength gain (about 80 per cent of its ultimate strength within 24 hours), and, from the 1950s in the UK, its use became quite widespread for precast components in buildings and bridges. In the early 1970s, a number of roofs made with HAC concrete collapsed. A major investigation by the Building Research Establishment (BRE) established that under normal service conditions, the compressive strength could be halved from the 1-day value over a long period of time. Basically, the hydrate products are chemically unstable and convert to a more stable form. This results in a more porous material, leading to any embedded metal being more vulnerable to corrosion. If HAC is suspected in a structure, a primary test method must be the taking of cores for strength measurement. The structural use of HAC was effectively banned in UK Codes in the early 1970s. Further details of HAC may be found in reference [4.9].

This brief review of cements is provided to make it clear that the source and type of cement should figure prominently in the investigative/diagnostic stage of any assessment, particularly if some form of chemical attack is suspected. While there is now a more rational approach to cement types in place (Table 4.5), the past history is both chequered and variable.

4.3.3.2 Aggregates

Aggregates generally make up at least 75 per cent of the volume of concrete, so it is not surprising that their quality is important. Sources of suitable aggregates for concrete making are many and varied, coming from a wide range of rock types with varying mineral compositions. These sources have been well researched over the years (see reference [4.9], for example), and desirable properties were established and set out in Standards, backed by a comprehensive portfolio of relevant test methods. This is an integral part of the subtleties of mix design, with strength, soundness, texture, size and shape being some of the key properties to ensure the required performance for both fresh and hardened concrete.

Aggregates can be natural or crushed, quarried or sea dredged. More recently, there is the growing source of recycled aggregates. Some aggregates may have less-desirable properties such as swelling or shrinking. In the context of this book, it is necessary to distinguish between properties which may lead to 'conventional/acceptable' defects and those which may have a more active role in a specific aggressive action.

Putting aside the importance of aggregate properties in abrasion resistance, possibly the key issue is that of 'impurities'. There may be organic impurities, clay, pyrites, or, more serious in durability terms, salts, sulfides and chlorides. There may also be the question of the minerals present in certain rock types. The alkali-silica reaction [ASR] cannot take place without the presence of a critical amount of reactive silica, for example.

In diagnostic terms, investigating the characteristics of the aggregates is probably less critical, compared with the case for cements, unless there are reasons to believe that the likelihood of impurities is high. There are exceptions to this, ASR being the most obvious one.

4.3.3.3 Concrete

In the context of diagnosis and assessment, Sections 4.3.3.1 and 4.3.3.2 briefly covered material considerations, for cement and aggregates, respectively. The intention here is to pickup other constituents, water and admixtures – while consolidating some of the lessons learned from feedback and Section 2.3, in particular.

4.3.3.3.1 WATER

Water here means water in the concrete mix for reasons of hydration and workability, with an additional role of providing adequate curing. As may be deduced from Table 2.2, water is also crucial in providing transport mechanisms for aggressive actions and in taking part in virtually all the resulting reactions; that aspect is dealt with later in Section 4.4. There is little that can be said about the quality of water used for concrete making. For very many years, the single word 'potable' has been used to represent the required quality. Standards exist in amplification of that, banning certain impurities and setting maximum levels for others that are deemed to be acceptable.

Water content and water/cement ratios (w/c) are perceived as key parameters in concrete mix design for strength. In general, the lower the w/c ratio, the greater the strength. More water is provided than is needed for hydration, for workability (consistency) reasons, i.e. to ensure ease of placement, irrespective of the method used. The process of hydration leads to the formation of cement hydrate gel around the particles of cement, and water may be trapped in pores or capillary passages. Some of this water may be used in subsequent hydration, leaving pockets of water vapour. Early on, these pockets or capillaries are interconnected, and may remain so, if too much water is present, if hydration is not allowed to develop over a sufficient period – or, indeed, if the degree of compaction is inadequate.

The above simplistic description is given to show that there is a strong link between the water provided in the mix, for whatever reason, and the operations on site in terms of placing, compacting and curing. The effect can be one of variable quality in terms of density, strength and permeability.

Figure 4.5, due to Neville [4.9], shows a relationship between density and concrete strength. Plainly, the level of compaction is a factor here – something that is ill-defined in specifications and standards ('the concrete shall be adequately compacted', not being unusual). The reduction in strength with density can be attributed mainly to the presence of voids in the concrete. However, there is also an interaction between compaction and water content, which should be considered if the existence of voids is to be minimised.

Figure 4.5 relates density to strength. There is the further consideration that the presence of voids will make it easier for aggressive actions to penetrate deeper into the concrete, particularly if these pores are interconnected.

Air, water and dissolved salts can all be involved in this process, and, as was noted from the feedback in Chapter 2, this is a major factor in reducing durability. Under field conditions, there is no general relationship between strength and permeability or porosity; there are just too many variables

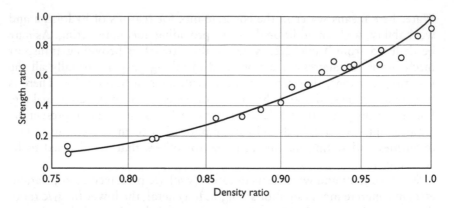

Figure 4.5 Relationship between strength and density ratios [4.9] (i.e. 1.0 = strength to be expected if full compaction achieved)

involved. If high permeability is suspected, possibly based on observation and on construction records, the only solution is to resort to testing.

Finally, there is the role of water in curing. In the industry, opinions have varied over the years as to the importance of curing, and, for certain types of cement, it is on the critical path and often difficult to do effectively – which means that it is not always done. Curing is an aid to the process of hydration, and the amount required depends on the type of cement and the w/c ratio. In the context of concrete permeability, it has a key role to play in blocking pores and continuous capillaries. Evidence of the curing regime from construction records is helpful at the investigative stage.

4.3.3.3.2 ADMIXTURES

The use of admixtures is very much on the increase in recent years, since they can provide technical and economic benefits, both during construction and in service. Fifty years ago, little was known about possible side effects of admixtures (the use of calcium chloride, as an accelerator, is the classic example of this). The situation now is much improved, even as the scope and types have increased, although it is probably still fair to say that the precise actions involved are not always fully understood, and performance can sometimes be variable if the correct dosage rate is not achieved.

A Cement Admixtures Association exists, which provides detailed information, both on what is currently available and on what may have been used in the past. There are also numerous handbooks and texts, e.g. references

[4.9] and [4.11]. In the context of diagnosis and assessment, the durability characteristics of the concrete are important; it is advisable to know exactly what is used in the concrete and its likely effects. The principal types of admixtures are:

Accelerators. Basically, these speed up early strength development by assisting the hydration of the calcium silicates.

Retarders. Here, the objective is to slow down the hardening of the cement paste, a useful role for special finishes or where continuous massive pours are involved. They are also useful in hot weather or where the concrete has to be transported long distances.

Water reducers/superplasticisers. This is probably the most interesting group in assessment, since it relates to water in the concrete and its possible influence on permeability. This group could be used in one of the two ways:

(i) to improve workability for a given w/c ratio;
(ii) to reduce w/c ratio for a given workability.

In (i), the concrete strength is not affected. In (ii), the resulting concrete is stronger for a particular cement content; alternatively, the cement content may be reduced and still provide the required strength. It may be seen that the most practical benefits occur during the construction stage, but the flowing nature of the concrete will be beneficial in minimising the number and size of the pores in the concrete, thus improving durability. This concept has moved on, with the introduction of self-compacting concrete.

In investigative terms, discovering whether or not an admixture has been used is valid for two reasons:

1. are there any possible adverse effects, which may be contributing to the observed deterioration;
2. if a plasticiser was used, exactly what was it, and what was the purpose of its use, e.g. reducing cement content. The use of a plasticiser may also be reassuring, in terms of a dense impermeable concrete having been provided.

Waterproofers. The purpose here is to prevent or limit the penetration of water into concrete. The action may be by blocking the pores or by lining them making the concrete hydrophobic. This could be a factor in establishing the internal moisture content.

Durability inhibitors. This is a general term to cover special cases such as corrosion inhibitors or the use of air entrainment to increase frost resistance.

4.3.3.4   Reinforcement and prestressing steel

Reinforcement and prestressing steel are manufactured to specifications contained in national and international standards, which tend to be modified or updated from time to time. It is therefore important to establish exactly what type of reinforcement is present, particularly for older structures. The basis of design may also be relevant, i.e. what stress–strain characteristics were assumed and what limiting stresses were used in design, compared with the actual loading history during the structure's life time.

Ferrous metals corrode, and this is probably the major deterioration mechanism for structural concrete. The nature and extent of the corrosion process will be critical, including whether or not the corrosion is general or local (pitting corrosion), plus some indication of the corrosion rate. Engineers will especially be concerned with any reduction in cross-sectional area and the possible effect of the expansive corrosion products on bond and anchorage. There is also some evidence for certain types of steel that corrosion can affect ductility. Here, there is no substitute for extracting samples from the structure and testing them.

4.3.4   Corrosion

Normally, reinforcement is protected by the highly alkaline nature of the surrounding concrete. This protection can be violated in one of two ways:

– by carbonation of the concrete
– by chloride ingress

The processes involved are fairly well understood, at least under controlled conditions, but there are many variables involved on site, which can affect the mechanisms of corrosion. This should be borne firmly in mind, when attempting to use the models that have been evolved, in diagnosing and predicting the progression and effects of corrosion in individual cases.

This section provides only a basic outline of the corrosion process, with diagnosis and assessment very much in mind. For greater understanding, the reader is referred to the literature; a selection is contained in references [4.1, 4.9, 4.10 and 4.12–4.15].

4.3.4.1   Carbonation

The pH level in concrete is normally about 12–13, permitting a protective oxide layer to form on the surface of the reinforcement. Carbon dioxide from the atmosphere can diffuse into the concrete, and, in the presence of moisture, can react with calcium hydroxide to form calcium carbonate. In time, the alkalinity of the concrete is reduced to a pH value of about 9–10;

then passivation is destroyed, and the reinforcement becomes vulnerable to corrosion, if sufficient moisture and oxygen are also present.

The carbonation process is progressive from the outer surface of the concrete as CO_2 penetrates through the carbonated concrete to the next layer. While the carbonated front may be irregular, due to local differences in the characteristics of the concrete and internal moisture conditions, its leading edge is very steep, with the transition in pH from 12–13 to 9–10 occurring in a few millimetres. The greatest rate of carbonation occurs at a relative humidity (RH) of about 50–70 per cent. In drier conditions, there is insufficient moisture present for the reaction to take place; in wetter conditions, the water can act as a barrier to the passage of the CO_2. Moisture conditions over the full in-service life of the structure therefore crucially affect the rate of carbonation.

4.3.4.2 Chloride ingress

Chloride ingress into concrete is complex, and may involve more than one process over long periods of time, depending as always on the characteristics of the concrete and the prevailing moisture conditions. The dominant process is normally assumed to be diffusion, but, especially where wetting and drying cycles are involved, capillary absorption – a faster process – can also occur. Here corrosion can occur even in highly alkaline conditions, with chlorides from the pore electrolyte in the concrete causing localised breakdowns of the protective oxide film.

In addition to the efficacy of the penetration process of the chlorides through the concrete, three other factors are significant.

1) The build up of the chloride concentration on the surface of the concrete. This depends on the sources of the chlorides and the continuity of their supply. It also depends on the local environment at the surface, dictated by the location and orientation of the structural element, e.g. some surfaces are directly washed by driving rain or run-off while others are not. Alternate wetting and drying cycles are also significant here.
2) Chlorides can be chemically or physically bound to cement minerals and hydration products. The critical feature then is the concentration of the unbound free chlorides which actually reach the reinforcement.
3) When these free chlorides do reach the steel, a critical amount has to accumulate before corrosion can start. This is known as the threshold level, and is not a fixed single value, but depends on:

    • type of cement, fineness, amount of C_3A, amount of gypsum, presence of pfa or slag, etc.;
    • w/c ratio;
    • moisture content, and its variation;

- oxygen availability;
- type of steel, surface roughness and condition;
- variations in the chloride profile through the concrete cover, and whether or not the chlorides reach the reinforcement uniformly along its length or at discrete locations.

The same concrete may therefore have very different threshold levels in different situations, and this has been found in practice [4.16]. In assessment work, the threshold level may be a key parameter in decision-making. A conservative approach would be to assume a value equal to that given in Codes for mixing water (0.4 per cent of cement weight) but values in excess of 1 per cent of cement weight have been measured in practice [4.16].

The modelling processes that have been developed for chloride-induced corrosion are largely based on apparent diffusion coefficients and measured (or assumed) values for surface chlorides [4.12–4.15]. Their main purpose lies in the development of more rigorous design for new construction. They are also useful, in support of measurements and diagnosis, in assessment work, but the above brief notes have been produced to indicate that this has to be done with some care. Generalisations are difficult; the corrosion process is largely structure specific.

4.3.4.3   Notes on the corrosion process

The mechanism of corrosion is electrochemical in nature, with some areas of the reinforcement becoming positively charged and others negatively charged (anodes and cathodes respectively) and the pore solution in the concrete acting as the electrolyte. The result is the production of ferrous hydroxide, which converts to rust in the presence of oxygen and water. This corrosion product has a volume greater than that of the parent bar, thus creating internal stresses. When these exceed the tensile strength of the concrete, cracking occurs along the line of the reinforcement, leading to possible spalling or delamination.

It is these physical effects of corrosion, together with loss of rebar cross-section and bond or even anchorage, that will be the main concern of the assessor. These aspects will be dealt with later in Chapter 5.

However, there are earlier aspects of the corrosion process, which the assessor needs to be aware of, and the purpose of this section is to draw attention to these. This will be done without going into the subtleties of the corrosion mechanism itself; details of that can be found in the literature [4.12–4.15]. The simple two-phase model for the corrosion process, shown in Figure 4.6, was originally proposed by Tuutti [4.17].

This model is helpful in assessment work, even though it may not truly reflect the complexities of what is actually happening in individual cases. For example, the corrosion rate is unlikely to be constant over a longish period of time. Nor is the transition between Initiation and Propagation likely to be so clearly defined, and the two straight lines of the model may translate into a curve, whose shape may be different for different locations in a structural element, even if these are quite close together. Nor does corrosion start uniformly around the full circumference of the bars, but at the face closest to the exposed surface or locally at cracks.

The helpfulness comes in defining the nature and scale of the investigative phase, depending on the owner's criteria for intervention and remedial action. For example, he may wish to intervene before point A is reached, in which case there is a strong emphasis on defining either the location of the carbonation front with some precision, or the chloride level at the reinforcement. On the other hand, if he allows for some corrosion to occur, without reaching a critical level, there will be more emphasis on determining corrosion rate and the nature of the corrosion cells that are present. In this latter case, 'damage' will require closer definition, in terms of percentage loss of rebar cross-sections and the occurrence of corrosion-induced cracking.

While the previous paragraphs and Figure 4.6 might be indicative of a diagnostic strategy, this has to be applied with both care and perspective. Mostly, the corrosion process proceeds by the formation of numerous microcells. The nature and intensity of these can vary, depending on the relative sizes of the anodes and cathodes and their location. Where chlorides are involved, macrocells can also develop, depending on the resistivity of the concrete and the uniformity of the penetration of the chlorides through the concrete. The galvanic current in macro-cell action can lead to local pitting rather than general corrosion, which is potentially more serious in structural terms.

In summary, and referring to Figure 4.6, it is unlikely that absolute answers will be obtained, no matter how detailed the investigation. Some form of risk assessment is usually necessary. This may be made at the level of

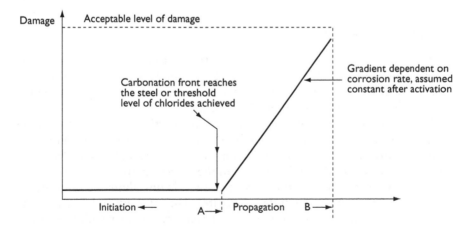

Figure 4.6 Tuutti model for corrosion [4.17]

scientific investigation, in interpreting the data from the many tests that are now available. However, ultimately, the concern lies with structural safety and serviceability, and it is generally prudent, as indicated in Figure 4.1, to undertake some calculations to take account of structural sensitivity to potential damage and the levels of redundancy in the design concept.

4.3.5   Sulfate attack

Chemical attack generally occurs via micro-structural change and the decomposition of the products of hydration. Chemical reactions occur between the constituents of the concrete and the penetrating action, and therefore involve the transport of ions, gas and moisture. The presence of water is crucial for most chemical actions to be significant.

Conventional sulfate attack has been recognised for many years and dealt with by defining classes of aggressivity and recommending concrete specifications for each, in terms of ingredients and mix proportions. The situation has become more complicated recently by the discovery of 'new' forms of sulfate attack, and greater use of brownfield sites, thus increasing the risk of a wider range of aggressive chemicals and disturbed impurities, which may increase sulfate content adjacent to concrete structures. This has led to a re-classification of aggressive chemical environments with corresponding material specifications [4.18].

In this section, only brief details are given to distinguish between the different types of sulfate attack and where they are most likely to occur. In any book on assessment, the focus must be more on effects than on causes, and with sulfate attack, all forms lead to expansion forces in the concrete. In each case, references will be given, permitting more fundamental studies to be made (e.g. reference [4.9], as a base and for reviewing the state-of-the-art internationally).

4.3.5.1   'Conventional' sulfate attack

Sulfates in the ground are usually of natural origin, the more important being those of calcium, sodium and magnesium. This means that the most likely locations for sulfate attack are in concretes from approximately ground level to a depth of about 10 metres.

The extent of the damage and the rate of the reaction are dependent on the nature and concentration of the sulfates, the mobility of the groundwater and the porosity of the concrete. In simple terms, the sulfate ions from the ground react with calcium hydroxide in the cement matrix to form calcium sulfate (gypsum). This in turn reacts with calcium aluminate hydrate to form calcium-aluminate hydrate (ettringite). The intensity and extent of the reaction therefore depend on the type of cement, together with the nature of the sulfates in the soil and the mobility of the groundwater.

The physical effects of sulfate attack may be summarised as:

- possible disintegration of the concrete due to expansion;
- softening of the concrete;
- an increase in moisture ingress, possibly containing aggressive salts.

These, in turn, could affect structural capacity, in terms of:

- loss of effective cross-sectional area of the concrete;
- a reduction in mechanical properties (strength, stiffness and bond/anchorage characteristics);
- a reduction in cover to the reinforcement, which, together with increased moisture ingress, would increase the risk of corrosion.

In assessment terms, the nature and extent of both the physical and structural effects would be determined during the investigative phase as a priority. However, the precise type of sulfate attack also requires to be studied, since this will affect the severity of the physical effects.

4.3.5.2 The thaumasite form of sulfate attack (TSA)

This form of attack is dealt with in detail in references [4.18, 4.19], the latter produced as a result of concerns regarding bridge foundations in the West of England. These involved buried concrete elements, containing carbonate aggregates, exposed to cold sulfate-bearing wet soils and groundwater. All the bridge foundations were exposed to large volumes of clay backfill, in which oxidation of pyrite had greatly enhanced sulfate levels. Reference [4.19] concluded that the number of structures potentially at risk to TSA in the UK was small. However, there is a need to be aware of the possibility, since the physical effects are potentially more serious than for conventional sulfate attack.

TSA is different from conventional sulfate attack, since it attacks the calcium silicate hydrates in the hardened concrete, rather than the calcium aluminate hydrates. The calcium silicate hydrates provide the main binding agent in Portland cements, and their conversion to thaumasite weakens the concrete, and, in advanced cases, the cement paste matrix is eventually reduced to a mushy white mass.

For thaumasite to form, the essentials are:

- a source of sulfates or sulfides in the ground;
- a source of calcium silicate hydrate from the cement;
- the presence of mobile groundwater;
- the presence of carbonate, generally in the concrete aggregates;
- low temperatures, generally below $15°C$.

The last two items are what distinguishes TSA from conventional sulfate attack, i.e. leading to the formation of thaumasite rather than gypsum or ettringite. The severity of the attack can also be influenced by secondary factors, such as:

- type and quantity of cement;
- concrete quality;
- changes in ground chemistry and water regime, arising from construction;
- type, depth and geometry of the buried concrete.

TSA in its early stages can be identified via laboratory analytical techniques. Advanced TSA can be identified visually, typically by a white pulpy mass at the surface of the buried concrete.

The physical and structural effects are basically the same as those identified in Section 4.3.5.1. The major difference is that the concrete could potentially become much softer; indeed, reference [4.19] advocates that concrete affected by TSA should be disregarded in strength calculations.

4.3.5.3 Delayed ettringite formation (DEF)

The mineral ettringite is commonly formed at early ages in Portland cement concrete cured at ambient temperatures; under these conditions, it is not damaging to concrete. However, if temperatures are high during the hydration process (either in large pours or in elements cured by internally applied heat), ettringite formation can be delayed, and its gradual formation in the cooled set concrete can lead to expansion and cracking. Opinions vary on what the critical temperature is, but it is of the order of 70°C, and the risk of expansion will increase the longer this temperature is maintained in the early hydration phase.

For ettringite to form, water must again be present, and its presence leads to an increase in the total solid volume of the cement and can be an expansive process. Time is important here. Initially, there is little or no expansion for a period of anything between 2 and 20 years – hence the name Delayed Ettringite Formation (DEF). While the processes involved are not fully understood, such expansion has occurred in a limited number of cases in the UK.

The dominant physical effect is likely to be cracking of the concrete due to expansion, leading to a reduction in mechanical properties. As with all expansive actions of this type, the structural effects will depend on the size and shape of the concrete element and the presence of internal or external restraints.

Further details regarding DEF may be obtained from reference [4.20] and its associated references.

4.3.6 Alkali–silica reaction (ASR)

Alkali–silica reaction is a reaction between the hydroxyl ions in the pore solution and certain forms of silica which may be present in the aggregate. It will only cause damage to the concrete when three factors are present:

- high alkali content in the concrete;
- a critical amount of reactive silica;
- sufficient moisture.

The reaction produces a gel which can swell and generate sufficient internal stresses within the concrete to cause cracking, which may be visually severe.

The causes of ASR are now well understood, and recommendations have existed in the UK since 1983 to minimise the risk of damage in new construction. It is unlikely therefore that structures built since then will suffer from ASR.

All aspects of ASR are now so well documented that it is not proposed to go into detail in this section, although the structural implications of the damage will be dealt with in Chapter 5. Instead, references are given under a series of headings as follows:

General background   References [4.21, 4.22]
Minimising the risk of ASR in new construction   References [4.23, 4.24]
Diagnosis of ASR   Reference [4.3]
Structural effects of ASR   References [4.25, 4.26]

Since ASR became a major concern first in the 1970s, the intensive R&D work carried out is a classic example of how to tackle a durability problem and provide authoritative guidance on all aspects. Science and engineering have worked in harmony, and we now have a good overall perspective. For assessment purposes, the emphasis is appropriately on the estimation of the expansion and its effect on mechanical and geometrical properties. In following through on structural implications, we now have sufficient knowledge to take account of both restraints and structural sensitivity. This will be brought out in Chapter 5, while demonstrating that the approach is capable of dealing with other aggressive actions.

4.3.7 Freeze–thaw action

When water freezes and turns into ice, there is an increase in volume of about 9 per cent. Therefore when water in concrete freezes, significant expansive stresses will only develop if the concrete is very close to full saturation, and these will occur in the depth to which frost has penetrated.

The result is internal mechanical damage, whose extent depends on the pore structure, generally involving micro-cracking. When and if wider cracks form on the surface of the concrete, the pattern is random and quite similar to that due to ASR, but with a tendency for some of the surface concrete to be dislodged more easily, in time. With poor quality concrete, this can lead to disintegration, as was the case with the Pipers Row car park described in Chapter 2.

Repeated cycles of freezing and thawing cause cumulative damage; cracking due to frost reaction in one cycle permits water in the following cycle to migrate to new locations and to cause further damage on freezing. Thus, expansive cracking which will have started fairly close to the surface can both spread over that surface and also penetrate deeper into the concrete.

As stated in Section 3.2.4.2, the practical effect of this internal damage is a reduction in mechanical properties. It has been found [4.27] that compressive strength is affected less than tensile strength and elastic modulus, most of all.

Freeze–thaw action can also lead to surface scaling. This is exacerbated if de-icing salts have been used. Exactly what happens here is complex and unclear. The salts can penetrate into the concrete pores, increasing the water content but lowering the freezing point. It is then possible for layering of frozen and unfrozen zones to occur, trapping water which may subsequently freeze. Whatever the mechanism, the effect is a loss of material at the surface, with the fines being disturbed first followed by the coarser aggregate. This is shown schematically in Figure 4.7, with a progressive

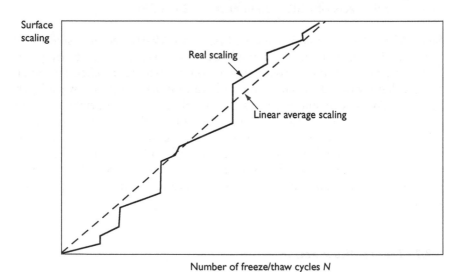

Figure 4.7 Salt-frost scaling curve with jumps caused by loosening of coarser aggregate particles

loss as the number of cycles increases. This effect reduces the cover to the reinforcement, thus increasing the risk of corrosion.

Freeze–thaw damage is treated very seriously in northern Europe for the obvious reasons of climate. However, the author believes that it is under-played in the UK, with observed damage being attributed to other expansive actions. A correct diagnosis is crucial, since the reductions in mechanical properties can often be very severe.

4.3.8 Leaching

Leaching occurs in structures exposed to pure water for significant periods; these might include dams, reservoirs, tanks and pipes. The water dissolves calcium-based components in the cement paste, which may be up to 65 per cent of cement weight in the concrete.

The effect can be worse in the presence of water pressure gradients across the structure, with high permeability concrete which allows water to flow through the structure. Leaching can also occur at cracks, with the water dissolving lime in the crack walls.

Leaching increases porosity and hence reduces strength and stiffness. A treatment of leaching is given in reference [4.28], with the emphasis on dams.

4.3.9 Acid attack

Portland cement concrete is highly alkaline and is not resistant to attack by strong acids or by compounds which may convert to acids. We are now entering a specialised area, in terms of the origins of the aggressive actions. Some acids may be used in industrial processes, e.g. in manufacturing fer-tiliser, which can also involve high temperatures.

Others may form from chain reactions of substances which, by them-selves, may be harmless to concrete, but may be converted to acids by bacterial action of some sort. Examples of this may be found in sewage, silage or manure handling systems. Therefore, we are faced with situations where the function of the structure will make it clear that some form of acid attack is a possibility. Some indication of the aggressivity of different acids to concrete may be obtained from Table 4.6. This is taken from refer-ence [4.1] and is based on guidance from the American Concrete Institute (ACI).

The creation of acids as the products of reactions of other substances is more difficult to illustrate because there are many possibilities. One

Table 4.6 Rate of attack by acids on concrete (Reprinted by permission of The Concrete Society. All rights reserved)

Rate of attack at ambient temperature	Inorganic (mineral) acids	Organic acids
Rapid	Hydrochloric Nitric Sulfuric	Acetic Formic Lactic
Moderate	Phosphoric	Tannic Humic
Slow	Carbonic	–
Negligible	–	Oxalic

possibility is sulfur compounds in sewage being reduced by bacteria to H_2S to form a weak mineral acid in water. Normally this is too weak to attack concrete, but, in closed systems such as sewers, it could condense on concrete above the waste water level, and, after oxidation, produce sulfuric acid. In general, concrete is not resistant to an acid environment with a pH value lower than about 4. The acid attacks the binding elements in the hydration products, dissolving the cement matrix.

Acid attack is covered in references [4.9, 4.29].

4.3.10 Other substances

Following on from Section 4.3.9, this is a general 'catch all' grouping of substances which may come into contact with concrete and possibly lead to disintegration.

Table 4.7 gives some examples. This is taken from reference [4.1], again based on guidance produced by the American Concrete Institute. It is not exhaustive, and the effects should be taken as indicative only, since they depend on the concentration of the products formed and on their pH level.

Note that disintegration is perceived as slow in the vast majority of cases, giving ample time to investigate both cause and effect.

4.4 Influence of the environment and micro-climate

In Table 4.4, water plays a central role in initiating deterioration mechanisms, with the exception of abrasion, and in controlling their intensity and rate of development. In general, it is the moisture state within the concrete which matters most, but this is very much influenced by the local external micro-climate. This can vary considerably in different locations

Table 4.7 Effect of some common substances on concrete (Reprinted by permission of The Concrete Society. All rights reserved [4.1])

Material	Comment on products formed	Effect
Ashes/cinders	If wet, sodium sulfate may leach out	Disintegrates concrete without adequate sulfate resistance
Beer	Fermentation products may contain acetic, carbonic, lactic or tannic acids	Disintegrates concrete slowly
Cider	Contains acetic acid (see Table 4.6)	Disintegrates concrete slowly
Coal	Sulfides leaching from damp coal may form sulfurous or sulfuric acid	Disintegrates concrete rapidly
Common salt		Not harmful to dry concrete Harmful to embedded steel in the presence of moisture
Creosote	Contains phenol	Disintegrates concrete slowly
Exhaust gases (diesel or petrol)	Form various acids in the presence of moisture	Disintegrates concrete slowly
Flue gases	Form various acids in the presence of moisture	Disintegrates concrete, slowly; temperature differentials may cause significant stresses
Fruit juices (and fermenting fruit)	Contain sugar and various acids	Disintegrates concrete slowly
Manure		Disintegrates concrete slowly
Milk		Not harmful but see 'Sour milk'
Peaty water	Contains humic acid	Disintegrates concrete slowly, if fatty oils are present
Petroleum oils		Disintegrates concrete slowly, if fatty oils are present
Silage	Contains a wide rage of acids	Disintegrates concrete slowly
Sour milk	Contains lactic acid (see Table 4.6)	Disintegrates concrete slowly
Sugars		Disintegrates concrete slowly
Urine		Attacks steel in porous or cracked concrete
Wine		Not harmful

and in different seasons of the year and cannot be ignored in assessment work.

The first general attempt, known to the author, to link micro-climate to deterioration processes, was in a CEB Bulletin [4.30]; the purpose of this was to provide input into the various models being developed by the CEB to represent various deterioration mechanisms, particularly corrosion. Table 4.8 shows the CEB approach.

The influence is expressed in terms of relative humidity RH, a useful concept in scientific terms, but possibly not representative of what happens under real conditions. CEB subsequently elaborated on this, in an attempt to develop a new approach to design for corrosion, taking account of local climate conditions [4.31]. In parallel, an approach has also been developed for design Codes and Standards, in defining exposure classes for specifying concrete quality and values for concrete cover to reinforcement. These exposure classes are now quite elaborate, as typified by Eurocode 2 [4.32], and the complementary British Standard to BS EN 206-1 [4.8].

So there is growing awareness of the importance of the local environment in durability design, at least as far as material specifications are concerned; the quality of concrete should therefore improve. The question then is: can

Table 4.8 Relationship between deterioration processes and relative humidity (Reproduced by permission of the International Federation for Structural Concrete)

Ambient relative humidity	Relative severity of deterioration process				
	Carbonation of concrete	Frost attack on concrete	Chemical attack on concrete	Risk of corrosion of steel	
				In carbonated concrete	In chloride-rich environment
Very low (<40%)	Slight	Insignificant	Insignificant	Insignificant	Insignificant*
Low – (40–60%)	High+	Insignificant	Insignificant	Slight	Slight*
Medium (60–80%)	Medium◊	Insignificant	Insignificant	High	High
High (80–98%)	Slight	Medium	Slight	Medium	Very high
Saturated (>98%)	Insignificant	High	High	Slight	Slight

Notes
* Corrosion risk in chloride-rich environments is also high when there are significant variations in humidity.
† For 40–50 percent relative humidity, carbonation is medium.
◊ For 60–70 percent relative humidity, carbonation is high.

these budding relationships between exposure classes and deterioration mechanisms be used in diagnosis and assessment, bearing in mind the increasing use of predictive models, which are highly dependent on accurate representative input for 'loads'.

To answer that, we need a perspective, which can come from Chapter 2. Figure 2.4 shows predictive moisture profiles. Table 2.2 provides a possible tabulation of environmental loads, from which must come the inputs for the predictive models. This is quite complex, while indicating the factors which will determine the accuracy of these inputs for individual structures. Some perception of the difficulties may then be obtained by looking at Figures 2.5–2.9, for different types of structure. Essentially, the local microclimate is structure-specific, and even then determining the influence of the different factors in Table 2.2 is not always straight-forward.

Looking at this in a little more detail, we find that the key factors are:

(i) local ambient conditions, and
(ii) moisture state inside the concrete.

Neither of these will be constant owing to diurnal and seasonal variations, and it is important to take that into account, if taking certain types of measurements on site at only one instant in time – corrosion rate, for example.

These key factors will also be modified by structure-specific issues, which can include:

(a) the interaction with wind, water and temperature;
(b) the influence of external boundary conditions, e.g. shape and texture of exposed elements; general structural or architectural detailing;
(c) the presence or otherwise of unplanned openings due to ineffective joints, or cracks caused by a lack of provision for movement;
(d) the provision made for water management, e.g. the efficiency and maintenance of drainage systems.

All of these issues can modify the nature of water in terms of vapour, run-off, driving rain or ponding. All can affect how much moisture can penetrate into the concrete and where that penetration can occur.

In spite of these difficulties, some effort should always be made at the investigative stage to assess the moisture state. This is important for three reasons:

1) As part of the diagnosis, in determining what is the dominant deterioration mechanism;
2) For making predictions on the likely future rate of deterioration;

3) Since water is a key component in most deterioration mechanisms, its elimination or control may be an attractive option in any proposed remedial plans.

How much effort is made in assessing moisture state may depend on what deterioration mechanism is dominant and on the scale and extent of the resulting damage. For example, for corrosion, it may well be worth looking at all the factors tabulated in Table 2.2, in order to make an evaluation, which is more quantitative than qualitative. In less critical situations, it may not be necessary to go into such detail. Nevertheless, beginning with the preliminary visual inspection, a start has to be made. One way of doing that is to build up the qualitative approach in Table 4.8. This has been attempted in reference [4.2] by the Concrete Bridge Development Group.

Table 4.9 is an extract from the CBDG report. The scope of Table 4.9 is wider than a simple concern with moisture state, and this has been built-in by a consideration of various types of defect for a range of deterioration mechanisms. Taking corrosion as an example from this Table, the environmental exposure classification is qualitative, but seen alongside the visual and test indicators, could point the way as to whether or not a more detailed analysis is necessary, based on the approach in Table 2.2. The CBDG report [4.2] follows up on Table 4.9 by giving guidance on the selection of tests to confirm a preliminary diagnosis. This subject is dealt with next in Section 4.5.

4.5   Inspection, testing and preliminary diagnosis

4.5.1   Background and approach

Referring to the basic action module in Figure 4.1, this section deals with the activities associated with levels 1 and 2. Essentially, it involves the gathering of information about the structure and on the extent and nature of the deterioration and its dominant cause.

It was noted in Chapter 3 that the formality of asset management systems can vary, but there is common ground in that different levels of inspection are always present, although the nature and frequency of these can vary. If, during one of these inspections, a change from the previous condition is noted – or significant deterioration detected – then a process comes into play to diagnose what is going wrong and the nature and extent of the irregularities.

While some preliminary tests may be carried out to augment the findings from the visual survey (Figure 4.1), the normal first step in the process is the desktop study to establish what is already known and to plan any testing regime, should this be considered necessary. Figure 4.1 suggests that

Table 4.9 Provisional diagnosis arising from routine inspection and testing (Reproduced with permission from the Concrete Bridge Development Group [4.2])

Visual and test indicators See Appendix D for diagrams and photos of cracking and defect references 1–25.

Inspection/test evidence	Reinforcement corrosion		Thermal cracking		Plastic cracking		Crazing	Drying shrink-age	Freeze–thaw damage	Alkali–silica reaction	Sulfate attack			Aggregate-related		Lime leaching	Fire damage	Impact damage	Load-induced
	Carbonation-induced	Chloride-induced	Early thermal	Progre-ssive	Shrink-age	Settle-ment					External	Internal (DEF)	Internal (thaum-asite)	Freeze–thaw Popouts	Pyrite Staining				
									Concrete Ducts										
1	3					1 [8]												1	
2		3	3	1															
3			3	3															
4	1				3			1		1	1								
5		1				3		1	1										
6					3	3													
7							3												
8								3	1	2	2								
9								2		2	2								1
10									2	2	2								2
11			1							2	1								
12	1								3	1									
13 [4]		1																	3
14 [4]								1											3
15 Spalling along reinforcement	3	3																	
16 Surface scaling									3		2								
17 Rust staining	3	3													2				
18 Exudations/surface deposits			1				1		1	1	2		2			3		1	
19 Stalactites																3			
20 Pinkish discolouration																	2		
21 Pop-outs over aggregate particles														2					
22 Spalling with no corrosion															2		2	2	
23 Impact damage																		3	

Cracking reference no. 1–14

Defect reference no. 15–23

Table 4.9 (Continued)

Inspection/test evidence	Reinf. corrosion: Carbonation-induced	Chloride-induced	Thermal: Early thermal	Progressive	Plastic: Shrinkage	Settlement	Crazing	Drying shrinkage	Freeze-thaw: Concrete	Ducts	Alkali-silica reaction	Sulfate: External	Internal (DEF)	Aggregate: Internal (thaumasite)	Freeze-thaw Popouts	Pyrite Staining	Lime leaching	Fire damage	Impact damage	Load-induced
24 Displacement of components											2	2							2	2
25 Surface decomposition								2	2			2	2							
Test indicators																				
Concrete delamination	3	3																		
Low cover [1]	3	2																		
Carbonation depth > cover	3																			
Chloride content moderate (M) [2]		2																		
Chloride content high (H) [2]		3																		
Environmental exposure																				
Exposed to rain	B		U	B	U	U	U	C	D	U	A	D	D	D	A	A	A	U	U	U
Sheltered from rain	A		U	B	U	U	U	C	D	U	C	D	D	D	D	C	C	U	U	U
Exposed to leakage/run-down	B		U	B	U	U	U	C	A	U	A	B	D	B	A	A	A	U	U	U
Exposed to spray	B		U	B	U	U	U	B	B	U	B	B	B	B	B	B	B	U	U	U
Buried	D		U	D	U	U	U	D	D	U	A	B	A[3]	A	D	A	B	U	U	U
Permanently water-saturated	D[5]		U	D	U	U	U	D	B[6]	U	A	A[7]	A	A[6]	B	A	A	U	U	U

Notes

[1] Definition of low cover must be based on an assessment of the exposure conditions and concrete quality.
[2] M = 0.3–1% Cl by mass of cement; H = > 1.0% Cl by mass of cement (based on Highways Agency guidance).
[3] Temperature is less than 15°C
[4] Cracking types 13 and 14 (load-induced) are the mostcommon. They are included to aid recognition, and to distinguish them from cracks due to durability problems. structural cracking may be more complex and require further investigation.
[5] Deterioration unlikely, even in saline conditions in tidal zone, as concrete can remain saturated.
[6] High risk in splash zone and due to wick action if hollow.
[6] If subject to freezing on a cyclical basis
[7] Depends on sulfate content of water. Worst case if high and replenished.
[8] If surface horizontal.
[9] If anchored zone.

Visual and text indicators
3 = Highly indicative
2 = May be indicative
1 = Rarely indicative, or may be associated

Environmental exposure
A = High risk
B = Moderate risk
C = Low risk
D = Determination unlikely
U = Deterioration unrelated to environmental exposure

the desktop study should include an evaluation of the design basis and structural sensitivity. This may be a key factor in establishing the nature and extent of any further testing – alongside the possible consequences of any failure occurring. Much depends on the outlook and approach and on the level of detailed information an owner feels he needs before taking a decision. There are no hard and fast rules about this, but merely some general principles which should be followed.

Basic categories of essential knowledge can be defined as follows:

(a) Design basis and construction records. If available information is insufficient, then some testing may be required to establish the as-built (undamaged) state. This might be augmented by sensitivity analyses on the possible effects of the deterioration, including coverage of geometrical and mechanical properties – sometimes checked via the testing of cores or by non-destructive test methods (NDT).

(b) Previous inspection and maintenance records – the structure's life story. What has been done to it? How effective has that been?

(c) The nature and extent of observed damage/defects. What are these? If cracking has occurred, what is its extent, format and possible cause? How widespread is it, and is it occurring at critical sections? Defects and their possible causes are covered in Sections 4.1 and 4.2. This is an important aspect in the early stages of the investigation, and a systematic approach is essential in noting, identifying and recording defects and damage. These records will form part of the structure's history in the future. A tabular approach, supported by photographs, sketches and notes, is the normal procedure – standardised as much as possible.

(d) Diagnosing the possible causes of the observed irregularities. Are they due to construction factors (Section 4.2) or is a deterioration mechanism involved (Section 4.3). Deducing the causes of any cracking is particularly important. In some cases the cause may be obvious, e.g. rust stains, but further investigation may be necessary to establish why corrosion has occurred, as well as its extent, in order to arrive at a reasonable figure for damage classification (level 3 in Figure 4.1). For other suspected aggressive actions in Table 4.4, such as sulfate attack or ASR, testing of the concrete may also be necessary, with a regime similar to that in Figure 4.2 for ASR. In all cases, a study of the local micro-climate is essential, for the reasons given in Section 4.4.

It is important at this stage to make a plan and to have clear objectives. What is to be measured? How detailed and precise should the results be? How are the results to be used in the decision-making process? What is really required to augment the information available from previous records and preliminary inspections? Focus is essential. Random or indiscriminate testing is not the way forward. The importance of this cannot be over-emphasised.

4.5.2   Test options

There are test methods available to cover the four categories of essential knowledge identified in Section 4.5.1. A listing is given in Table 4.10. This is not comprehensive, but gives some indication of the options available for seeking further information on the different properties being investigated. Some have been in use for many years, with widespread experience of their use. Some are relatively new, and novel techniques are coming onto the market, with greater capability and, apparently, with greater precision.

It is quite common to use more than one method in investigating a particular property. This may be simply as a cross-check, for example in anything to do with corrosion or corrosion rate. More commonly, it is to obtain a broader perspective, e.g. NDT methods are used to obtain a qualitative picture of concrete quality and strength in the structure as a whole to identify suspect areas from which cores may be taken or samples obtained, for detailed laboratory analysis.

Testing of this type is usually undertaken by test houses or specialist sub-contractors, who have detailed knowledge of their preferred methods, experience of their use and limitations, and expertise in interpreting the results. Many of the methods are covered by Standards, such as those from BSI, CEN or ASTM. Descriptions of the more common methods appear in the literature, e.g. references [4.1, 4.2, 4.29] and [4.33 – 4.35].

4.5.3   Test locations and sampling

The choice of test locations will be based on observations made during the preliminary survey, in studying the structure as a whole and identifying areas where the deterioration is most severe. NDT and near-surface tests can be helpful in this regard.

What is done in practice will depend on whether only one site visit is involved or the structure is being regularly monitored. It will also depend on reaching an acceptable compromise between available resources (time, money, labour) and the need to obtain results which are sufficiently accurate and representative.

An understanding of population, mean and standard deviation is important here, as is some judgement of the likely variability of the property under investigation. A particular difficulty is the measurement of any characteristic which is subject to variation in time and significantly influenced

Table 4.10 Relevant tests methods

Property under investigation	Test method/inspection technique
Concrete strength	− Taking and crushing cores − Near surface tests (pull-off; internal fracture; penetration resistance, etc.) − Rebound hammer
Concrete quality/uniformity	− Visual examination of cores − Ultrasonic pulse velocity (upv) − Impact echo tests − Radiographic examination − Petrographic examination − Chemical analysis
Location of reinforcement or prestressing steel	− Cover meter − Physical exposure − Pulse echo tests − Sub surface radar
Depth of carbonation	− Phenolphthalein testing − Petrographic analysis
Presence of chlorides	− Chemical analysis (surface, or drillings for profiles)
Presence of sulfates	− Site chemical tests, supported by laboratory analysis
Corrosion of reinforcement	− Electro-potential mapping − Resistivity measurements − Physical exposure
Corrosion rate	− Linear polarisation resistance − Galvanic current measurement
Cracking, spalling, delamination	− Visual inspection, mapping, photographs − Sounding surveys − Impact echo tests, upv, thermography, etc.
Surface permeability/absorption	− Surface absorption tests (ISAT) − Water and gas permeability tests
Moisture content	− Direct measurement on site or on samples
Alkali–silica reaction	− Petrography − Laboratory tests for free or restrained expansion
Frost action	− Petrography − Laboratory tests
Material properties	− Taking of samples for physical or chemical testing − Strength tests for concrete or reinforcement

by the local micro-climate – moisture content, for example, or corrosion rate. The precision of the test method, and its repeatability, is also a factor.

Sampling is a key feature in the development of a test regime. Plainly, the desire will be to obtain as much representative and reliable data as possible. In interpreting the data, however, a perspective is necessary in appreciating the limitations due to the inevitable compromise on testing regime.

4.5.4 Selection of tests to confirm a provisional diagnosis

The heading of this section implies that some indication of the likely cause of the deterioration has been obtained from the preliminary inspection. This is important in providing focus and in deciding what tests should be carried out.

An outline has already been given in Figure 4.2 of a recommended diagnosis for ASR. This is quite comprehensive and has been assembled with the objective of gaining as much representative and reliable data as possible – after the possibility of ASR has been identified at an early stage. Similar processes can be developed for other suspected deterioration mechanisms – for example, for the various facets of corrosion listed in Table 4.10.

Test house experience and expertise are important in doing this in individual cases, and the derivation of universally-acceptable rules is difficult. An attempt to provide this has been made in reference [4.2] for concrete bridges, and Table 4.11 is an extract from that report, which gives some indication of the type of thinking involved in developing a test programme. The approach here is to classify tests as Essential, Desirable, and Possibly Helpful, for each of the durability parameters. In this way, a realistic test programme can be built up for individual deterioration mechanisms, which may be similar to that in Figure 4.2 for ASR, while allowing for structural sensitivity and the perceived degree of damage. The principle is that the test plan should cover all the relevant categories of knowledge given in Section 4.5.1, to enable a reliable rating to be given at the Preliminary Assessment stage (level 3 in Figure 4.1).

4.6 Interpretation and evaluation: pre-preparation for preliminary assessment

Quite deliberately, no attempt is made in Section 4.5 to describe the individual test methods. Proper coverage would require a separate book. Some details can be found in the references to this chapter and in the literature in general; however, be mindful that test methods are constantly evolving and new ones are introduced.

Instead, the emphasis has been put on formulating a testing plan with clear objectives on what is to be measured, on the required level of detail,

Table 4.11 Selection of tests to confirm a provisional diagnosis (Reproduced with permission from the Concrete Bridge Development Group [4.2])

Durability parameter	Site tests																			Laboratory tests											
	Visual inspection	Crack width measurement	Coring over crack	Delamination survey	Covering measurement/rebar locator	Half-cell potential	Resistivity	Linear polarisation resistance	Carbonation	Rebound hammer	Autoclam permeation tests	Surface absorption (ISAT)	Radar	Breakout of concrete	Optical/digital inspection of cavities	Ultrasonic pulse velocity	Radiography	Pressure tests on ducts	Thermography	Chloride content	Sulfate content	X-ray diffraction	Cement content	Alkali content	Carbonation	Petrographic examination	Surface absorption	Capillary absorption	Permeability	Autoclam permeation tests	Expansion of cores
Reinforcement corrosion – carbonation-induced	3			3	3	3	2	2	3		1	1	1	2						3			1		3	2	1	1	1	1	
Reinforcement corrosion – chloride-induced	3			3	3	3	2	2	3		1	1	1	2						3			1		2	2	1	1	1	1	
Early thermal cracking	3	3	1		1																										
Progressive thermal cracking	3	3	1		1																										
Plastic shrinkage	3	3	3																												
Plastic settlement	3	3			2																										
Crazing	3	3																					1			2					
Drying shrinkage	3	3			1																2					3	1	1			
Freeze–thaw damage	3	3			1																2		1			2	1	1			
Alkali–silica reaction	3	3	3		1																		3	3		3					3

Table 4.11 (Continued)

Site tests

Durability parameter	Visual inspection	Crack width measurement	Coring over crack	Delamination survey	Covering measurement/rebar locator	Half-cell potential	Resistivity	Linear polarisation resistance	Carbonation	Rebound hammer	Autoclam permeation tests	Surface absorption (ISAT)	Radar	Breakout of concrete	Optical/digital inspection of cavities	Ultrasonic pulse velocity	Radiography	Pressure tests on ducts
Sulfate attack – external	3	3			1													
Sulfate attack – internal (DEF)	3	3	3		1													
Sulfate attack – internal (thaumasite)	3	3	3		2													
Aggregate freeze–thaw pop-outs	3																	
Pyrite staining	3																	
Lime leaching	3	1																
Prestressing tendon durability	3				1									3	3	1	3	3
Fire damage	3				3				3	3						2		
Impact damage	3				3					2				3				
Load-induced cracking	3	3		2	2					1						1		

Laboratory tests

Durability parameter	Thermography	Chloride content	Sulfate content	X-ray diffraction	Cement content	Alkali content	Carbonation	Petrographic examination	Surface absorption	Capillary absorption	Permeability	Autoclam permeation tests	Expansion of cores
Sulfate attack – external		2	3		1			3					
Sulfate attack – internal (DEF)		2	3		1			3					
Sulfate attack – internal (thaumasite)			3	3				3					
Aggregate freeze–thaw pop-outs								3					3
Pyrite staining								1					
Lime leaching								2					
Prestressing tendon durability													
Fire damage							3	3	2				
Impact damage							3	3					
Load-induced cracking					1			1					

Key: 3 = Essential, 2 = Desirable but not essential, 1 = May be helpful in particular cases

and on that detail being both representative and reliable. In doing that, it is also necessary to consider how the results are to be used, at the level 3 Preliminary Assessment stage in Figure 4.1. It is assumed that some form of damage classification will be the norm at this early stage, possibly associated with simple structural calculations and attempts to model the perceived dominant deterioration mechanism.

Chapter 5 of this book will use the SISD rating approach developed from the CONTECVET programme [4.14], [4.22], [4.27], but it was recognized in Chapter 3 that this is not the only approach. If, for example, a method similar to that in Table 3.2 for the Canadian Bridge Code is adopted, the inspection and test data will have to be evaluated rather differently. Conversely, if the simple Condition Index approach in Figures 3.9 and 3.10 are to be used, then pre-calibration will be necessary to link the numerical values to the condition description emerging from the inspection and testing phase. Additionally, several sets of indices will probably be necessary for serviceability and strength, possibly even for different action effects (bending, shear, bond, cracking, etc.).

The point being made is that condition rating is not just a simple matter of interpretation of test data – often difficult enough in itself, particularly for tests which do not give a direct measure of the parameter being investigated. An integral part of the testing plan has to be the performance factors which are important to the owner and what is acceptable to him in terms of minimum performance for each of these. This will obviously influence the decision on when to intervene, as well as the nature of that intervention. But, more important at this planning stage, it will dictate what is to be measured during the testing stage.

This important issue is discussed in Section 3.3.3, in association with Figure 3.12, with examples given of possible performance criteria. The vertical axis in Figure 3.12 is labeled 'Load capacity' and might involve any or all of the structure-specific performance requirements listed in Table 4.12. These are the ultimate concerns of the assessor, but, within a progressive assessment approach, are probably examined directly only at the Detailed Investigation stage (level 4 in Figure 4.1), should the Preliminary Assessment indicate the need to do so.

This means that the condition rating approach at level 3 has to be sufficiently robust to permit a sensible decision to be taken. This affects the scope of the test programme and has an influence on how the results are interpreted.

To illustrate this point for corrosion, Figure 3.12 is reproduced as Figure 4.8, but with a change in the vertical axis – instead of 'load capacity', there is 'corrosion-related performance'. For cases where corrosion is perceived as the main deterioration mechanism, Table 4.10 reveals that the parameters of interest would include the following, set out in three groups as follows:

Table 4.12 Structure-specific performance requirements during service life

1. Safety	Strength	Fire resistance
	Stiffness	Fatigue
	Stability	Resistance to exceptional loads (e.g. Impact, seismic effects)
2. Serviceability	Deformation	Vibration
	Displacement	Water-tightness
	Cracking	
3. Operation and function	Minimisation of downtime	
	Inspectability	
4. Durability	No unexpected or exceptional repair bills	
	No significant reduction in the performance requirements in 1–3 above.	

Group A   Element dimensions
            Steel percentage and detailing
            Cover to reinforcement
Group B   Rust stains
            Cracking – nature, location, width, spacing
            Delamination
            Loss of rebar cross-section
Group C   Depth of carbonation
            Presence of chlorides

            –   at the surface
            –   at the reinforcement

            Corrosion rate
            Local micro-climates
            Diffusion characteristics of the concrete

Groups A and B are concerned with the nature of the structural element and the present state with respect to damage and deterioration, i.e. the gathering of information directly relating to the performance requirements in Table 4.12, as far as present condition is concerned. Group C relates to the mechanism itself, and, if damage has already occurred, would be focused mainly on predicting the gradient of the performance line in Figure 4.8.

However, there are frequent cases where the conditions for corrosion to occur are suspected, but it may not have started. The owner may wish to adopt one of a number of alternative management strategies, which might include intervention:

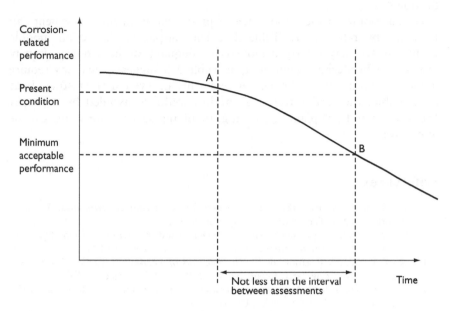

Figure 4.8 Corrosion-related performance versus time

- before carbonation or chloride fronts have reached the reinforcement;
- when the fronts have just reached the reinforcement;
- when corrosion has just started;
- when cracking due to corrosion has just started; or
- when the loss of rebar cross-section has reached x **per cent.**

Each of these could be set as minimum performance requirements on the vertical axis of Figure 4.8. They may be seen as relatively small differences, but the time to intervention between each can be substantial – and of economic benefit to the owner, provided of course that none of the performance requirements in Table 4.12 are at risk. The adoption of such a strategy would have a major influence on the nature and scope of the test programme. Corrosion has been used to illustrate the importance of pre-preparation prior to the preliminary assessment – the need to have a plan, to determine what is to be measured and how the data is to be used. The prime interest will always be with the factors listed in Table 4.12, but there may be merit in constructing diagrams like Figure 4.8 for the parameters that are central to the dominant deterioration mechanism. This can be done for all the mechanisms given in

Table 4.4, while taking note of the effects of each of these as given in Section 4.3.

A final point on the importance of pre-preparation for assessment can be made by reference to Table 3.8. The output of the action module in Figure 4.1 may clearly demonstrate adequacy or inadequacy. It may also give a borderline result, and, as Table 3.8 suggests, this may require a move forward to a Detailed Assessment, or even more testing. If at all possible, the further-testing scenario should be avoided by a proper focus and good pre-planning; testing, involving several site visits, can be expensive.

References

4.1 The Concrete Society. *Diagnosis of deterioration in concrete structures*. Technical Report 54. 2000. The Concrete Society, Camberley, UK.

4.2 Concrete Bridge Development Group (CBDG). *Guide to testing and monitoring the durability of concrete structures*. Technical Guide No 2. 2002. CBDG, Camberley, UK.

4.3 British Cement Association (BCA). *The diagnosis of alkali–silica reaction*. 2nd Edition. Report of a Working Party. Report No 45.02 BCA, 1992. Camberley, UK.

4.4 The Concrete Society. *Non-structural cracks in concrete*. Technical Report 22. 1982. The Concrete Society, Camberley, UK. December.

4.5 The Concrete Society. *The relevance of cracking in concrete to corrosion of reinforcement*. Technical Report 44. 1995. The Concrete Society, Camberley, UK.

4.6 Buenfeld N.R. *Advances in predicting the deterioration of reinforced concrete*. The 7th Sir Frederick Lea Memorial Lecture. The Institute of Concrete Technology (ICT) Year Book: 2004–2005. pp. 23–38. 2005. ICT, Crowthorne,UK.

4.7 The Concrete Society. *Changes in the properties of Ordinary Portland Cement and their effects on concrete*. Concrete Society Technical Report 29. 1987. The Concrete Society, Camberley, UK.

4.8 British Standards Institution. BS 8500. Concrete – Complementary British Standard to BS EN 206-1. BSI, London, UK. 2002.

4.9 Neville A.M. *Properties of concrete*. 4th Edition. Longman Group Ltd, Harlow, UK. 1995.

4.10 The Concrete Society. (Author: P.B. Bamforth). *Enhancing reinforced concrete durability*. Technical Report 61. 2004. The Concrete Society, Crowthorne, UK.

4.11 Hewlett P.C. (Editor). *Cement admixtures, use and applications*. Longman, Harlow, UK. 1988. p. 166.

4.12 Broomfield J.P. *Corrosion of steel in concrete*. Understanding, investigation and repair. E.&F.N. Spon, London. 1996. p. 240.

4.13 Page C.L., Bamforth P.B. and Figg J.W. (Editors). *Corrosion of reinforcement in concrete construction*. Proceedings of the 4th International Symposium, Cambridge, UK. July 1996. Royal Society of Chemistry, UK.

4.14 CONTECVET (BRITE-EURAM PROJECT IN30902D). *A validated user manual for assessing corrosion-affected concrete structures*. 2000. Available from the British Cement Association, Camberley, UK.

4.15 Building Research Establishment (BRE). *Corrosion of steel in concrete: durability of reinforced concrete structures*. Digest 444 (4 Parts). BRE, Garston, UK. 2000.

4.16 Pettersen K. Chloride threshold value of the corrosion rate in reinforced concrete. *Proceedings of an International Conference on corrosion and corrosion protection of steel in concrete.* University of Sheffield, UK, 1994. Also available the Swedish Cement and Concrete Research Institute (CBI), Stockholm, Sweden.

4.17 Tuutti K. *Corrosion of steel in concrete.* BCI Forskning: Research Report to 4.82. 1982. Swedish Cement and Concrete Research Institute (CBI), Stockholm, Sweden.

4.18 Building Research Establishment (BRE). *Concrete in aggressive ground.* BRE Special Digest No 1 (4 Parts) 2nd Edition. BRE, Garston, UK. 2005.

4.19 Department of Environment, Transport and the Regions (DETR). *The thaumasite form of sulfate attack: risks, diagnosis, remedial works and guidance on new construction.* 1999. Report of the Thaumasite Expert Group. DETR, London, UK. January.

4.20 Building Research Establishment (BRE). Delayed ettringite formation: insitu concrete. BRE. Information Paper IP11/01. 2001.

4.21 Hobbs D.W. *Alkali silica reaction in concrete.* Thomas Telford Ltd, London, UK. 1988.

4.22 Clark L.A. *Critical review of the structural implications of the alkali silica reaction in concrete.* Transport Research Laboratory (TRL). Contractor Report 169. 1989. TRL, Crowthorne, UK.

4.23 Building Research Establishment (BRE). *Alkali silica reaction in concrete.* BRE Digest 330 (4 Parts). BRE, Garston, UK. 1997.

4.24 The Concrete Society. *Alkali silica reaction.* Minimising the risk of damage to concrete. Technical Report 30 (3rd Edition). 1999. The Concrete Society, Camberley, UK.

4.25 Institution of Structural Engineers. *Structural effects of alkali silica reaction.* Technical guidance on the appraisal of existing structures (2nd Edition). IStructE, London, UK. 1992.

4.26 CONTECVET IN30902I. A validated Users Manual for assessing the residual service life of concrete structures affected by ASR. A deliverable from an EC Innovation project. Available from the British Cement Association, Camberley, UK. 2000.

4.27 CONTECVET IN30902I. A validated Users Manual for assessing the residual service life of concrete structures affected by frost action. A deliverable from an EC Innovation project. Available from the British Cement Association, Camberley, UK. 2000.

4.28 Fagerlund G. Leaching of concrete. The leaching process: extrapolation to deterioration. *Effects on structural stability.* Report TVBM-3091. Division of Building Materials, Lund Institute of Technology, Sweden. 2005.

4.29 Bijen J. *Durability of engineering structures.* Design, repair and maintenance. Woodhead Publishing Ltd, Cambridge, UK. 2003. p. 262.

4.30 Comite Euro-International Du Beton (CEB). Durable concrete structures. CEB Information Bulletin 183. Thomas Telford Ltd, London, UK. 1992. p. 112.

4.31 Comite Euro-International Du Beton (CEB). New approach to durability design – an example for carbonation-induced corrosion. CEB Information Bulletin 238. CEB, Lausanne, Switzerland, and Thomas Telford Ltd, London, UK. 1997.

4.32 British Standards Institution (BSI). BS EN 1992-1-1. Eurocode 2: design of concrete structures – Part 1: general rules and rules for buildings. BSI, London, UK. 2004.

4.33 fib. Management, maintenance and strengthening of concrete structures. Bulletin 17. *fib,* Lausanne, Switzerland. 2002. p. 174.

4.34 Kay E.A. Assessment and renovation of concrete structures. Longman Scientific & Technical. Harlow, UK. 1992. p. 224.

4.35 Bungey J.H. and millard S.G. Testing of concrete in structures. Blackie Academic & Professional, Glasgow, UK. 3rd Edition 1995. p. 286.

Chapter 5

Preliminary structural assessment

5.1 Introduction

This chapter is concerned with level 3 assessment in the action-module shown in Figure 4.1 (repeated for convenience as Figure 5.1). The objective is to develop a formal numerical approach to damage classification, as proposed in the CONTECVET Manuals [5.1]. However, as may be clearly seen from the figure, this should not be done in splendid isolation, without taking some account of structural implications; this is essential in providing a perspective, and as an aid to decision-making.

The interpretation of data from inspections and selected tests is also an important issue. Little has been said about that in Chapter 4, since, as indicated in Section 4.5.2, this is a specialist subject to be undertaken by experienced experts with detailed knowledge; however, some coverage is necessary here, in terms of the principles involved, with emphasis on assessing the structural implications alongside the formality of the selected damage classification method.

Therefore, the sequence in this chapter is as follows:

5.2 Interpretation of test data
5.3 Engineering perspective in support of damage classification
5.4 Preliminary assessment; general principles and procedures
5.5 Preliminary assessment for ASR
5.6 Preliminary assessment for frost action
5.7 Preliminary assessment for corrosion
5.8 The nature and timing of intervention
     References

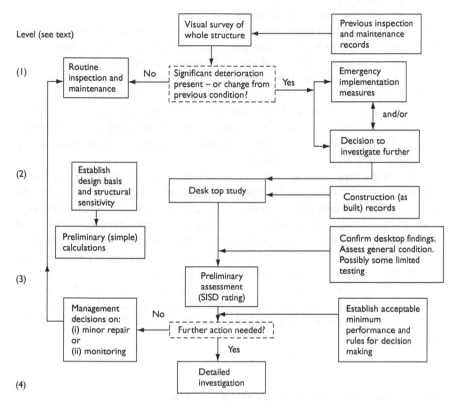

Level (see text)

(1)

(2)

(3)

(4)

Figure 5.1 Action module – flow diagram for progressive screening (Reprinted by permission of The Concrete Society. All rights reserved [5.1])

5.2 Interpretation of test data

5.2.1 Introduction

Aspects of this were dealt with in Sections 4.5 and 4.6, and this section will deal only with some basic principles, while creating an overall perspective. As indicated in Table 4.10, the range of available and relevant tests is large and varied, and interpretation requires experience and expertise. Guidance is available, and references that the author has found helpful include the following:

General guidance – References [5.2, 5.3, 5.4]
ASR – References [5.1, 5.5, 5.6, 5.7]
Frost action – Reference [5.8]
Cracking – References [5.9, 5.10]

This represents a personal selection. There are numerous additional references, including most of those listed in Chapter 4.

For most of the scientific tests in Table 4.10, particularly those that are laboratory based, procedures are given in Standards on the methodology, with some guidance on the handling of the data and on precision. Generally, there is less certainty on what constitutes a critical result, in terms of its significance in the context of structural deterioration under less controlled site conditions. For many of the properties under investigation in the table, more than one test method is available – some simple and physical, others more technical – and, if resources and timescale permit, it is prudent not to rely solely on a single test method.

Other tests give only an indirect measure of the property being investigated – e.g. the rebound hammer for concrete strength – and here the interpretation is especially important, in terms of both relevant and representative (the sampling issue). Any tests where moisture condition and micro-climate are significant parameters require special attention in interpretation, since these conditions can vary considerably both locally within the structure and seasonally over the structure's life.

Practical common sense and judgement are essential, together with a standard and consistent approach, while attempting correlation, or even calibration, with results from similar investigations which appear in the literature.

5.2.2   Collection and classification of data

The need for photographs, sketches and standard descriptions of the observed deterioration was emphasised in Chapter 4. These are valuable both in themselves and in helping to identify critical areas where more in-depth investigation may be desirable.

Figure 5.2, taken from reference [5.3], emphasises this point. This is a sketch of a bridge pier, showing different forms of deterioration and varying micro-climate conditions; it also indicates where drilling samples were taken in this particular case, and the location of previous repairs. This type of information is invaluable in providing a perspective to the investigation as a whole.

Cracking is another area where a systematic approach to the collection and classification of data is essential. General treatments of cracking and its significance are given in references [5.9] and [5.10]. Table 5.1, taken from reference [5.2] and based on original work by RILEM Committee TC 104, is an attempt to classify defects in terms of damage ratings. In reference [5.3], the approach is slightly different, as typified by Table 5.2; here, crack spacing is dealt with more formally, but without any attempt at assigning damage ratings.

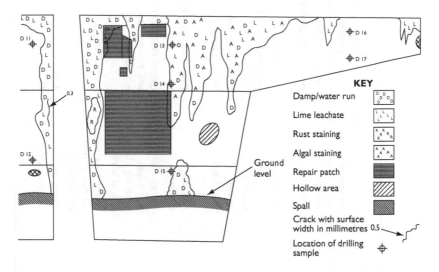

Figure 5.2 Example of a sketch record of visual inspection of bridge pier (Reproduced with permission from the Concrete Bridge Development Group [5.3])

This latter approach has merit, since assigning a damage rating purely on the basis of crack width, without reference to the probable cause of the cracking, may not give a true picture of structural significance. In addition, as indicated in Section 4.4 (Tables 4.8 and 4.9), the local micro-climate has to be considered in evaluating deterioration rates. Structurally, the location of the cracking with respect to critically loaded sections must also be considered. Tables 5.1 and 5.2 are indicative only; the real need is to have a systematic approach and to evaluate the results in a wider (structural) context.

5.2.3 Overview of data interpretation

Table 4.10 gives a listing of test methods, targeted at quantifying different properties. Table 4.11 attempts to prioritise these, with some classed as essential, and others as 'used less frequently, if at all'. The maxim 'do not take any measurements unless you know how to use the results' really applies to this selection process.

To appreciate the needs in interpretation, it is necessary to make a distinction between site and laboratory tests, and to understand how the test equipment measures the property involved. To illustrate this, Table 5.3 contains a selection of tests from Table 4.10, indicating which ones are site- or laboratory-based, and whether or not a direct or indirect quantitative reading is obtained or the results are merely qualitative. To be clear:

Table 5.1 Examples of classification of defects [5.2]

Damage	Damage rating				
	1 (Very slight)	2 (Slight)	3 (Moderate)	4 (Severe)	5 (Very severe)
Cracks in prestressed concrete due to overloading	Width <0.05 mm	Width 0.05–0.1 mm	Width 0.1–0.3 mm	Width 0.3–1 mm	Width 1–3 mm with some spalling
Cracks in reinforced concrete due to overloading	Width <0.1 mm	Width 0.1–0.3 mm	Width 0.3–1 mm	Width 1–3 mm with some spalling	Width >5 mm with widespread spalling
Cracks in unreinforced concrete	Width <1 mm	Width 1–10 mm	Width 10–20 mm	Width 20–25 mm	Width >25 mm with spalling
Shrinkage or settlement cracks	Single small crack	Several small cracks	Many small cracks	Few large cracks	Many large cracks
Effects of reinforcement corrosion	Barely noticeable	Light rust stains	Heavy rust stains	Heavy rust stains and cracking along line of bars	Heavy rust stains and spalling along line of bars
Pop-outs	Barely noticeable	Noticeable	Holes up to 10 mm in diameter	Holes between 10 and 50 mm in diameter	Holes >50 mm in diameter
Spalling	Barely noticeable	Clearly noticeable	Larger than coarse aggregate	Areas up to 150 mm across	Areas larger than 150 mm

Table 5.2 Description and classification of cracks: (a) crack width and
(b) crack spacing (Reproduced with permission from the
Concrete Bridge Development Group [5.3])

(a) Crack width

Surface crack width (mm)	Description
<0.1	Very fine
0.1–0.3	Fine
0.3–1	Moderately wide
1–5	Wide
5–10	Very wide
>10	Extremely wide

(b) Crack spacing

Distance between adjacent cracks (m)	Description
<0.025	Extremely closely spaced
0.025–0.1	Very closely spaced
0.1–0.25	Closely spaced
0.25–0.5	Moderately spaced
0.5–1	Widely spaced
1–5	Very widely spaced
>5	Extremely widely spaced
Single	Isolated

Direct   means the equipment gives a direct value for the property being
   investigated;

Indirect   indicates the property is determined indirectly, either by calibration or via an assumed relationship; and

Qualitative   means the test will not yield quantitative results.

Each category presents different problems of interpretation. The majority of
direct tests are laboratory based, and carried out via established procedures,
often found in Standards, with guidance given on interpretation of the
results. The problem comes in relating these results to the structure on site.
Sampling then becomes important (Section 4.5), in terms of number and
location, and in the case of categories 4 and 5 in Table 5.3, in terms of
time as well, since most of the measurements are sensitive to micro-climate
and moisture conditions – and these can vary both within the structure and
seasonally.

Indirect and qualitative tests are generally site based, and again sampling
is important, but the key feature is the calibration of the results with the

Table 5.3 Different types of interpretation required for a selection of test methods

Property under investigation	Test	Interpretation of results	Site or laboratory test	Comments
1. Concrete strength	Cores	Direct	Site/laboratory	Laboratory tests are generally carried out on cores, lumps or dust samples.
	Near-surface tests	Indirect	Site	
	Rebound hammer	Qualitative	Site	
2. Concrete quality	Visual examination (structure; lumps etc.)	Qualitative	Site/laboratory	
	Ultrasonic pulse velocity	Indirect	Site	
	Chemical analysis	Direct	Laboratory	
	Petrographic examination	Direct/indirect	Laboratory	
	Expansion of cores	Direct	Laboratory	
3. Reinforcement issues	Rebar or tendon location	Indirect	Site	Indirect measurement is via cover meters Direct measurements by breaking out the concrete locally at selected locations.
	Depth of cover	Indirect/direct	Site	
4. Moisture/gas penetration/diffusion	Local micro-climate, and nature of surface moisture (run-off, spray etc.)	Direct	Site	In this category, laboratory tests are carried out on samples removed from the structure. Sampling is therefore a major issue.
	Concrete porosity	Direct	Laboratory	
	Initial surface absorption	Direct	Site	
	Water permeability	Direct	Laboratory	
	Gas permeability	Direct	Laboratory	
5. Corrosion of reinforcement	Carbonation depth	Direct	Site laboratory	
	Chloride content/profiles	Direct	Laboratory	
	Half-cell potential and potential mapping	Indirect	Site	
	Corrosion rate	Indirect	Site	
	Resistivity	Indirect	Site	

Table 5.4 Relationship between half-cell potential and
risk of corrosion [5.2]

Half-cell potential	Risk of corrosion
Less negative than – 200 mV	5%
–200 to –350 mV	50%
More negative than –350 mV	95%

Table 5.5 Relationship between resistivity and corrosion rate [5.2]

Resistivity (kΩcm)	Likely corrosion rate
<3	Very high
5–10	High
10–20	Low
>20	Negligible

required property. In general, these tests have developed substantially over the past 20–30 years, as experience has been gained, but the classification of the results for interpretation purposes can still be fairly general. Guidance on this for the tests listed in Tables 4.10 and 5.3 is given in references [5.2] and [5.3], but to illustrate the point, Table 5.4, taken from reference [5.2], shows the generally accepted relationship between half-cell potential readings and risk of corrosion. Similarly, Table 5.5 gives a relationship between resistivity and likely corrosion rate. Using this type of information in individual cases is not easy and requires expertise, knowing that both properties can be much affected by moisture conditions. The definition of risk of corrosion and corrosion rate in Tables 5.4 and 5.5 respectively is also fairly general, giving a ball-park figure and acting mainly as an indicator of what additional investigations may be necessary, relating to relevant properties in Table 5.3.

In the above, emphasis has been put on reinforcement corrosion, probably the most significant deterioration mechanism. In trying to quantify this in individual structures, the range of necessary test extends beyond category 5 in Table 5.3. The importance of moisture state and its variability has already been stressed and cannot be over-emphasised. Similarly, if chlorides are present or carbonation suspected, but corrosion not yet started, then attempts may be made to model penetration profiles, in which case the characteristics of the concrete become important, in terms of permitting diffusion or capillary suction. It is important in all of this to have a clear purpose in both test selection and data interpretation.

5.3   Engineering perspective in support of damage classification

5.3.1   Introduction

In moving towards a preliminary structure assessment (level 3 in Figure 5.1), the action-module includes items such as:

(a) preliminary (simple) calculations
(b) established design basis and structural sensitivity
(c) examination of construction records
(d) review of data from previous inspection and maintenance records

Exactly what is done here will depend on what information is available, the owners future management strategy and the preferred approach of the assessors. Should construction records be unavailable, and information from previous inspections sparse, more in-depth investigation may be necessary via items (a) and (b) above, supported by results from preliminary inspections and test data.

While preliminary calculations are regarded as essential, some assessors may prefer to indulge in more rigorous analysis, both for modelling the deterioration process and for structural assessment – procedures which are highly dependent on the quality and reliability of the input. Whatever may be the preferred approach in arriving at the rating based on damage classification, an awareness of the structural issues is important, as an aid to deciding on what action to take in moving on from level 3 in Figure 5.1.

This section gives brief details of some of the issues and possible approaches.

5.3.2   Overall picture

Table 4.12 lists the main structural performance requirements during the service life of a structure. Table 5.6 is an elaboration of that, with a stronger emphasis on the action effects normally considered in design for strength calculations; this is a move towards identifying the mechanical and sectional properties which might be affected by deterioration.

In design, acceptable margins or safety factors are provided, mostly using standard methods provided in Codes. Structurally, in assessment, the objective is to determine that these have not reduced to an unacceptable level, due to the effects of deterioration.

Once materials begin to deteriorate, resistance is provided by the structure itself, and there are various facets of that whose importance may vary for different types of structure. Table 5.7 makes some suggestions on this, giving ratings on four facets for a range of structural types – with rating 1 being the most important, and rating 4, the least.

Table 5.6 Structure-specific performance requirements during service life

Safety	Stability	Overall Local Progressive collapse Fire	
	Strength	Rupture of sections	flexural tension flexural compression direct compression direct tension shear (transverse) shear (in plane) punching shear torsion local bearing (and spalling) bond anchorage (+ combinations of these)
		Elements and connections	buckling impact fatigue
Serviceability		Stiffness/articulation Cracking	 corrosion (local and general)
		Deflection Deformation	appearance expansion dilation overstress
Function		Vibration/dynamic response Weather/water tightness Movement Appearance	

Table 5.7 Relative importance of different facets of design as contributors to structural protection, for different types of structures (1 = most important; 4 = least important)

Structures	Design concept	Design details	Materials	Structural details
Residential	2	1	4	3
Commercial	2	1	4	3
Public/social	2	1	4	3
Industrial	2	1	3	4
MSCP	2	1	3	4
Marine	1	3	1	4
Bridge	1	2	3	4
Tunnel	4	3	1	2

Design details are rated marginally higher than design concept for buildings in general. Marine structures are heavily reliant on both design concept and materials; this is because of their basic nature (often massive) and their unrelenting exposure to aggressive actions. Bridges, on the other hand, are more reliant on design concept, in terms of the provisions frequently made to protect against direct exposure. The approach is simplistic, but an attempt to identify the more important facets of design in pure protection terms.

Design concept is important in other ways, particularly in determining structural sensitivity. Some indication of this is given in Table 5.8 for some typical structural systems. As an example of this, the failure of the Pipers Row car park described in Section 2.5 was initiated because frost action led to the destruction of the tension chord in the internal resistance mechanism to punching shear; the extent of the failure was dictated by the lack of tie reinforcement to prevent progressive collapse.

The consequences of failure will always loom large in developing an overall structural perspective. Structural redundancy will be a major factor in sensitivity terms, and so also the existence of internal or external restraints due to movement or strain – an example of this latter issue will emerge in Section 5.5 when dealing with expansion due to ASR.

5.3.3 Structural analysis in perspective

Figure 5.1 strongly suggests the need for calculations at an early stage in assessment. Because of the major advances made in computer-based analytical methods in a relatively short time, it is tempting to use these powerful methods (capable of dealing with non-linearity, secondary effects etc.) to solve the assessment problem, assuming that the greater rigour gives 'better' values for capacities compared with those obtained by simple linear methods. This indeed is the basis for tackling bridges in the UK as described in Section 3.2.2.1. A perspective is necessary since the capacity predictions are model-dependent: different capacities are obtained from different models, e.g. linear and non-linear Finite Element Analysis; even for a given model, different answers can emerge from mesh refinements.

In assessment, the use of these complex methods has to be made with some care, particularly if the output is judged against stress-related acceptance criteria. In doing this, it is not uncommon to demonstrate 'over-stress' and take that as failure, hence indicating an urgent need for remedial action. The maximum load that can be carried is not necessarily governed by the first critical section reaching its apparent capacity. Further,

Table 5.8 Sensitivity to deterioration of typical structural systems

Structural system	Structural concept	Critical features	Protection systems
Simply-supported beams	• Only one hinge required for failure		• Materials and cover • Links
Continuous beam	• Three hinges required for failure	• Ponding over supports • Local micro-climate	• Redundancy • Materials and cover • Links
Simply-supported slab	• Yield lines required for failure		• Failure over large area required • Materials and cover • Effect of deterioration averaged out over width due to alternative load paths around deterioration
Continuous slab	• Yield line system incorporating supports and mid-span required for failure	• Local micro-climate • Joint details	• Redundancy • Failure over large area required • Materials and cover • Effect of deterioration averaged out over width due to alternative load paths around deterioration

Table 5.8 (Continued)

Structural system	Structural concept	Critical features	Protection systems
Flat slab	• Yield line system incorporating supports and mid-span required for failure • Or punching shear at column support	• Ponding over columns • Joint details	• Materials and cover • Tie reinforcement to prevent progressive collapse
Column	• Buckling • Crushing	• Loss of cover reduces section and increases possibility of reinforcement buckling	• Materials and cover • Links
Corbel	• Bearing resistance	• Ponding on top surface • No alternative load path	• Materials and cover • Links

estimates of capacity calculated in this way are dependent on the assumptions made and the relevance of the input. Particular examples of this include:

(a) the assumed boundary conditions – usually taken as being either pinned or fixed, but generally somewhere in between and almost impossible to estimate;
(b) the capability for redistribution – much dependent on detailing, which is not always amenable to mathematical precision; and
(c) the actual mechanical and sectional properties as affected by deterioration.

Complex structural analysis does not figure strongly at level 3 assessment, except possibly for owners coping with large populations of structures on an ongoing basis. Its use comes more into play should a detailed investigation be deemed necessary. What is needed however is an awareness of structural sensitivity and how this might be affected by deterioration.

Table 5.8 gives some indication of this. For collapse to occur, a failure mechanism has to form, whose nature depends on the type of element and its redundancy for slab-like structures. Table 5.8 puts a strong emphasis on yield lines, and yield-line analysis is frequently used to obtain an upper bound solution. However, there are other issues raised in Table 5.8, which are more directly related to the effects of deterioration; e.g. for columns or walls, a severe loss of cover could convert a short column into a slender one, with the additional risk of the longitudinal reinforcement buckling, should the links be poorly anchored.

Taking this aspect further, consider the case of a beam as shown in Figure 5.3. Should the expansive action of the corrosion products lead to cracking and spalling, there is a risk of loss of bond for the tension reinforcement (Figure 5.3(a)). This alters the strain that can develop in the tension reinforcement (Figure 5.3(b)), but, provided the anchorage of the tension steel remains intact, an alternative strut-and-tie mechanism can develop (Figure 5.3(c)). In fact, the effect of spalling on bond strength is difficult to determine except by physical testing, but there are data available which clearly indicate the major contribution to bond strength when stirrups are present [5.11].

Figure 5.3 is a simple example of extreme deterioration leading to a possible alternative resistance mechanism. Other possibilities exist for the different structural systems in Table 5.8, but the most common case is where the deterioration reduces the performance requirements in Table 5.6. The objective then is to quantify that reduction, and aspects of this will emerge in Chapter 6 covering detailed investigation, carried out by Webster [5.11], based on the CONTECVET manuals and on the extensive review of published experimental data on deterioration.

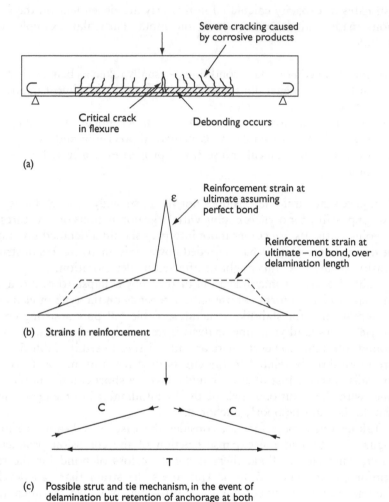

(a)

(b)   Strains in reinforcement

(c)   Possible strut and tie mechanism, in the event of
       delamination but retention of anchorage at both
       supports

Figure 5.3 Effect of significant loss of bond of tension reinforcement on the magnitude
and type of failure

Nevertheless at the preliminary assessment stage (level 3), it is useful to
have some idea of what material and structural parameters can be affected
by different deterioration mechanisms. An indication of this is given in
Table 5.9, for the more common deterioration mechanisms.

Taking this simple perspective a little further, it may also be helpful
to develop a preliminary risk assessment. Based on experience and on
a review of the literature, Table 5.10 puts forward a proposal which
is indicative rather than quantitative. Safety hazard and frequency are

Table 5.9 Important parameters for structural assessment

Deterioration mechanism	Compressive strength	Tensile strength	Bond strength	Rebar area	Anchorage	Laps	Links	Restraint
Early age								
Plastic settlement								
Plastic shrinkage								
Early age thermal movement								
Environmental								
Long-term drying shrinkage								
Creep								
Moisture-sensitive aggregates								
Long-term thermal movements								
Freeze–thaw: Surface scaling			✓		✓	✓	✓	
Freeze–thaw: Internal damage	✓	✓	✓		✓	✓	✓	
Chemical								
Corrosion: Carbonation-induced			✓	✓	✓	✓	✓	
Corrosion: Chloride-induced			✓	✓	✓	✓	✓	
Alkali–silica Reaction (ASR)	✓	✓	✓		✓	✓	✓	✓
Sulfate attack			✓		✓	✓		
Delayed ettringite formation (DEF)	✓	✓	✓		✓	✓	✓	✓
Structural								
Flexure	✓		✓	✓	✓	✓	✓	
Shear	✓	✓	✓	✓	✓	✓	✓	
Axial	✓			✓	✓	✓	✓	

expressed in Table 5.10 as numbers between 1 and 5, and the priority score is obtained by multiplying the two together. Indicative definitions of frequency and safety hazard are given in Tables 5.11 and 5.12 respectively.

It is stressed that Table 5.10 is empirical and subjective to some extent. It is also stressed that the prime concern is with safety. With those qualifications in mind, the priority score values are of some interest. Corrosion is clearly identified as the biggest risk, followed by freeze–thaw action, with the other mechanisms being relatively minor in terms of safety.

Table 5.10 Preliminary risk assessment of deterioration mechanisms

Deterioration mechanism	Possible consequences	Safety hazard	Frequency	Priority score
Early age				
Plastic settlement	i) Loss of bond between top bars and concrete			
	ii) Provides route into cover zone for moisture, oxygen and chloride	I	4	4
Plastic shrinkage	i) Provides route into cover zone for moisture, oxygen and chlorides	I	4	4
Early age thermal movement	i) Provides route into cover zone for moisture, oxygen and chlorides	I	4	4
Environmental				
Long-term drying shrinkage	i) Contraction and/or deflection	2	3*	6
Creep	i) Contraction and/or deflection	2	2	4
Moisture-sensitive aggregates	i) Contraction and/or deflection	2	3	6
Long-term thermal movement	i) Opening and closing of existing cracks and/or joints, and possibly new cracks	2	3	6
Freeze–thaw: Surface scaling	i) Loss of cover leading to reduction in durability			
	ii) Reduction in bond strength	3	4*	12
Freeze–thaw: Internal damage	i) Loss of cohesion within concrete			
	ii) Loss of concrete compressive strength	4	4*	16
	iii) Loss of concrete tensile strength (approximately twice the loss of compressive strength)			
	iv) Reduction in bond strength			
	v) Loss of load-carrying capacity			
Chemical				
Corrosion: Carbonation-Induced	i) Spalling of cover concrete	4	5	20
	ii) Loss of reinforcement cross-section			
	iii) Loss of bond strength			
	iv) Loss of load-carrying capacity			
Corrosion: Chloride-induced	i) Spalling of concrete cover	4	5	20
	ii) Loss of reinforcement cross-section			

Mechanism	Effects			
	iii) Loss of bond strength			
	iv) Loss of load-carrying capacity			
Alkali-silica reaction (ASR)	i) Loss of tensile and compressive strengths of concrete ii) Chemical prestress iii) Little or no loss of load-carrying capacity that cannot be explained by loss of concrete strength	2	2	4
Sulfate attack	i) Loss of cover ii) Loss of strength in small sections	2	2	4
Delayed ettringite formation (DEF)	i) Possible reduction in freeze–thaw resistance ii) Substantial reductions in elastic modulus and tensile and compressive strength	4	1	4

Structural

Mechanism	Effects			
Flexure	i) Lower load-carrying capacity than designed	4	1	4
Shear	i) Lower load-carrying capacity than designed	5	1	5
Axial	i) Lower load-carrying capacity than designed	5	1	5

Table 5.11 Indicative definitions of frequency in Table 5.10

Frequency	Definition
1	Extremely rare
2	Very rare
3	Rare
3*	Rare, but present in localised areas
4	Uncommon
4*	Uncommon, but present in localised areas
5	Common

Table 5.12 Definitions of safety hazard in Table 5.10

Severity	Definition
1	Mild
2	Moderate
3	Severe
4	Very severe
5	Extreme

5.3.4   Structural sensitivity versus hidden strengths

In Sections 5.3.2 and 5.3.3, the emphasis is on structural sensitivity arising from the basic nature of the structure (redundancy, etc.) and how capacities may be reduced by deterioration. There is another side to this particular coin – the possibility of hidden strengths. These may arise from alternative load-bearing mechanisms, compared with those assumed in design, which may be brought out by different models. Figure 5.3 is a simple example of this.

Almost always, an assessment of the actual imposed loading should be considered; if this is less then the design value, then a reduced design resistance may be acceptable while retaining the same margins and safety factors, i.e. $R_d/S_d \geq 1$ becomes $R_a/S_a \geq 1$, with the assessed values of Resistance (R_a) and Imposed Load (S_a) both being less than the original design values.

However, the more general situation is a re-evaluation of R_d, in terms of either:

(a) its magnitude, evaluated by alternative means,
   or
(b) the resistance being provided by alternative load paths.

Aspects of this will be dealt with in Chapter 6; the perspective at the level 3 stage in the action-module (Figure 5.1) is that it is not all doom and gloom in structural terms. Reference [5.12] covers some of the ground, in giving guidance on how and when to use alternative structural models for deteriorating concrete bridges.

5.3.5   Engineering perspective of corrosion

The brief perspective of structural aspects, provided in Sections 5.3.2–5.3.4, involves looking at the appraisal process rather differently compared with the flow diagram in Figure 5.1. The starting point is the structure itself, its nature and how it might perform, assuming some level of deterioration yet to be quantified by inspection and testing.

Traditional structural appraisal process	Stages in a structural appraisal and the questions that each stage should address	Alternative structural appraisal process
	Inspection • What is causing the structure to deteriorate? • How much have the materials deteriorayed to date? and • How much will they deteriorate in the future? Assessment • What is the load-carrying capacity of the structure now? and • What will it be in the future? • How fast will it deteriorate? Management • How safe is the structure now? and • How safe will it remain? • What remedial measures will be required? and • How much will they cost? • What happens if no remedial measures are carried out?	

Figure 5.4 Balancing the traditional and alternative appraisal process, to develop a structural perspective and to maintain focus (Reprinted by permission of The Concrete Society. All rights reserved)

This is illustrated in Figure 5.4, in terms of following the alternative appraisal process on the right hand side of the figure. The most fundamental questions about safety and economy are posed under the management process. The implication is that unless you know what questions to ask, the answers needed will not be obtained. So, a basic question is: how safe is the structure now and how long will it remains safe? This then gives the linear approach in Figure 5.1 much more focus.

A similar line of thinking can be developed to obtain a perspective of corrosion risk and corrosion rate, and their likely structural effects. Evaluating corrosion risk and rate in the field can be difficult. There is a great deal of variability due to local micro-climates and how these can vary in both short and long term – and there is the problem of practical interpretation as typified by Tables 5.4 and 5.5. Measurement has to be done with care, and interpretation involves both judgement and experience.

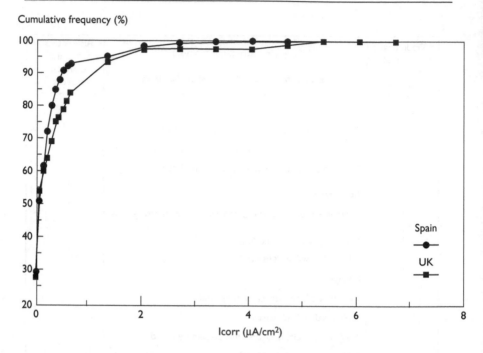

In view of this, a structural engineer may wish to obtain a range of estimates, based on different assumptions, of what corrosion might do structurally. During the CONTECVET project [5.13], corrosion-rate measurements were obtained on a variety of Spanish and UK structures. These are presented in Figure 5.5, in the form of a cumulative frequency diagram. In the UK, approximately 95 per cent of the structures had corrosion rates less than 1.0 µA/cm². Taking this value for first estimation purposes, we have:

$$I_{corr} = 1\mu A/cm^2 = 11.6 \ \mu m/year \quad i.e. \ 0.0116 \ mm/year$$
$$\Phi_t = \Phi_I - 0.023 \ I_{corr}.t \quad I_{corr} \ is \ in \ \mu A/cm^2$$
$$(0.023 \ converts \ this \ to \ mm/year)$$
$$\Phi_i = initial \ bar \ area$$
$$\Phi_t = bar \ area \ at \ time \ t$$
$$For \ a \ 20 \ mm \ \Phi \ bar, \ area = \mid \frac{\pi d^2}{4} \mid = \pi \times 100 = 314.29 mm^2$$

At 50 years, and $I_{corr} = 1 \ \mu A/cm^2$, loss of bar area $\approx$ 10 % (for 20 mm bar)
$$\approx 25\% \ (for \ 8 \ mm \ bar)$$

This type of thinking can lead to the type of chart shown in Figure 5.6 for a 20 mm Φ bar and a range of corrosion rates, assuming uniform

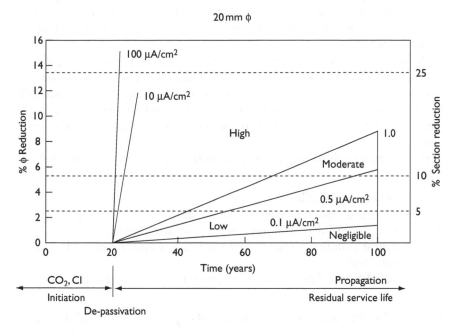

Figure 5.6 Possible design chart for estimating possible loss of rebar cross-section for a 20 mm bar

corrosion. As a minimum, this can give a perspective; with checks provided by reliable site measurements, it can move towards a more quantitative approach. Even greater refinement is feasible, particularly via spreadsheets [5.11], in making allowance for possible pitting corrosion or in feeding in a range of data for chloride concentrations either on the surface or at the level of the reinforcement.

5.4   Preliminary assessment; general principles and procedures

In this book, assessment is seen as progressive in the sense of going only as far as is necessary to arrive at a decision on what action, if any, to take. In general, the progression is linear as shown in Figure 5.1, with inputs coming in at different levels. On occasions, to obtain a structural perspective, aspects of the two alternatives shown in Figure 5.4 may come into play.

It may of course be feasible to reach a decision at either level 1 or level 2 in Figure 5.1, depending on what has been found from the initial survey, should the design basis be sound and the progress of the deterioration mechanism limited. Preliminary assessment is perceived as a formal qualitative step in the process, in standardising decision-making.

Preliminary assessment involves the derivation of a SISD rating, standing for Simplified Index of Structural Damage. It should not be too elaborate or complex, being essentially qualitative, but the structural perspective can be a factor in deciding on the level of detail, as can the extent of information available from previous records and preliminary testing.

In evaluating a SISD rating, the factors to be taken into account are:

1) Consequences of failure;
2) Environment, micro-climate, moisture state;
3) Current level of deterioration – nature, location, extent;
4) Structural detailing – sensitivity to strain/movement, caused by deterioration; and
5) Predicted future deterioration rate, based on (1)–(4) above.

It may also be necessary to consider the physical effect of the deterioration on section or mechanical properties, e.g. reduced concrete strength due to ASR or frost action, or loss of cover via spalling or delamination.

The nature of the deterioration mechanism is a factor in how this is done, e.g.

–   for ASR, the extent of the expansive action is central;
–   for frost action, the relative extent of internal mechanical damage and surface scaling;
–   for corrosion, environment aggressivity is crucial, as is corrosion rate. The rating must have an element which reflects corrosion damage (including whether it is general, or pitting is involved due to the nature of the corrosion cells).

Reinforcement detailing is another major influence on the SISD rating.

What we have, in effect, is a representative standard system, which accounts for all the above factors and is focused on decision-making at level 3 in Figure 5.1, having established minimum acceptable performance requirements. The methodology used here was developed during the CONTECVET project [5.1], [5.8], [5.13], and the best way to describe it is to follow the procedures in the three Manuals, for ASR, frost action and corrosion. This follows in Sections 5.5–5.7.

5.5   Preliminary assessment for alkali–silica reaction

5.5.1   Introduction

There is a great deal of published information about ASR, as indicated in Section 4.3.6, which also gives the key references for the UK on:

(a) minimising the risk of ASR in new construction;
(b) diagnosis; and
(c) structural effects of ASR.

These references are well established, and it is not proposed to dwell on them here, other than to make the point that they were the starting point for further development work undertaken during the CONTECVET project, with emphasis on structural assessment [5.1].

By the time a preliminary assessment is being considered, quite a lot will be known not only about the structure but also about the cause and effect of the deterioration. For ASR, much of that will come from the diagnosis process, in establishing that it is the primary mechanism. Figure 4.2 gives an outline of that process, which clearly shows that most of the key factors identified in Section 5.4 will have been treated; moisture state, crack patterns and widths, and especially, movements and displacements will all have been investigated.

Research clearly shows that the magnitude of the expansion due to ASR is the dominant factor in evaluating structural effects of the damage. This therefore is a key feature in deriving a SISD rating. It is also the basis for establishing other detrimental effects such as reductions in mechanical properties or the possibility of over-stress in heavily loaded structural elements. This determines the procedures detailed in the sections which follow.

5.5.2   Free or restrained expansion

Early guidance on ASR [5.3], [5.25] was based mainly on estimates of free expansion derived from tests on cores. The amount of expansion depended on the alkalis in the cement (type of cement, cement content), the type and amount of the reactive silica in the aggregate and the moisture state; even for cores of broadly similar mixes taken from similar structural elements, there was considerable variability. However, expansion in the structure is restrained, with that restraint being provided by the structural elements themselves (boundary conditions etc.) and/or by the presence of reinforcement.

In the CONTECVET project, SISD ratings were derived for both free and restrained expansion. It is in fact possible to estimate restrained expan-

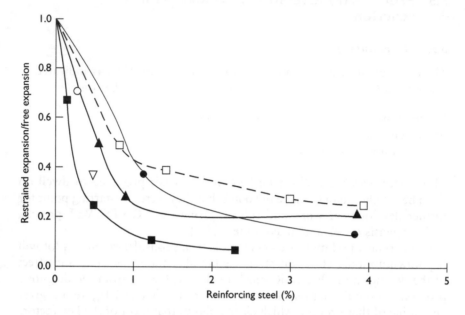

Figure 5.7 Typical experimental data showing the effect of restraint from reinforcement on expansion

sion using crack-width-summation methods [5.14]. However, in CON-TECVET [5.1] due account was also taken of published test data of the type shown in Figure 5.7, relating the ratio of restrained to free expansion to the reinforcement percentage, from 56 separate sources and covering a range of expansion rates. It may be seen that even quite a small steel percentage can significantly reduce expansion, but that the scatter of data is considerable.

In providing restraint, the reinforcement is stressed in tension, and the concrete parallel to the reinforcement is stressed in compression. Using this information, and based on data additional to that in Figure 5.7, a correction/conversion chart was produced [5.1]. This is shown in Figure 5.8 This enabled SISD ratings to be produced for both free and restrained expansion.

5.5.3 Residual mechanical properties

Expansion in the concrete due to the action of ASR can cause internal mechanical damage in the form of micro-cracking and hence reduce the strength and stiffness of the concrete. The magnitude of the reduction depends on the amount of expansion, which, in turn, can affect different properties in different ways.

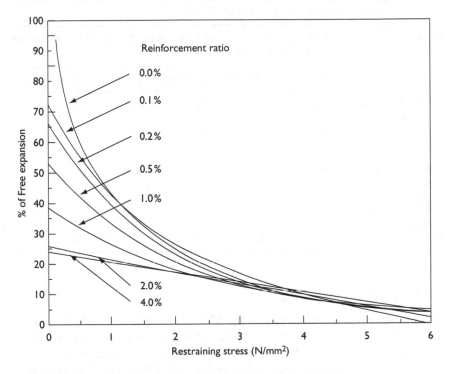

Figure 5.8 Proposed corrections for restraint (CONTECVET [5.1])

Best estimates of this can be made by taking cores from unaffected areas of the structure and using that as a base. Free or restrained expansion can then be estimated as indicated earlier, and an estimate made of the reduction in strength or stiffness by using the correction factors given in Table 5.13. This is taken from the report by the Institution of Structural Engineers [4.25] and relates residual mechanical properties to the 28-day compressive strength specified in the design. A better estimate can probably be obtained by deriving a value for the uni-axial strength from cores as suggested above. However, at this preliminary stage, we are looking at ball-park figures only, and, in that context, Table 5.13 is reasonable. For a detailed investigation (level 4 in Figure 5.1) where residual resistance to action effects are estimated, a more focused approach will be necessary.

Table 5.13 shows that compressive strength is the least affected as expansion increases, and elastic modulus the most. This is typical for this type of internal damage, and a similar picture will emerge in Section 5.6, for frost action.

Table 5.13 Lower bounds of residual mechanical properties for ASR-affected concretes based on specified 28-day strength [4.25]

Property	Percentage strength as compared with unaffected concrete for various amounts of free expansion				
	0.5 mm/m	1.0 mm/m	2.5 mm/m	5.0 mm/m	10.0 mm/m
Compressive cube strength	100	85	80	75	70
Uniaxial compressive strength	95	80	60	60	–
Tensile strength (appropriate to split and torsional tensile tests)	85	75	55	40	–
Elastic modulus	100	70	50	35	30

5.5.4 Structural detailing

Because of the expansive nature of the ASR, the severity rating SISD should make some allowance for the confining effect of reinforcement detailing. There are several ways of doing this (see Section 5.7), but, for ASR, it is proposed to follow the classification in CONTECVET [5.1], as proposed originally by the Institution of Structural Engineers [4.25].

Three classes of reinforcement detailing are recognised as shown in Table 5.14. These are illustrated in Figure 5.9(a) for walls and slabs and in Figure 5.9(b) for columns. Class 1 will rarely be found, and most structural elements will fall into class 2. Some walls and slabs may have no transverse steel, and will then be in class 3, the least effective in terms of restraining the expansive effects of ASR.

Even where transverse reinforcement is present, the anchorage needs to be checked. Figure 5.10 illustrates some possibilities for links. The most

Table 5.14 Reinforcement detailing classes for SISD rating

Class	Confinement of expansion	Typical example
1	High	3-dimensional cage of very well-anchored reinforcement
2	Medium	3-dimensional cage of well-anchored reinforcement
3	Low	2-dimensional cage of reinforcement in one or two faces; no through ties, no links or low over

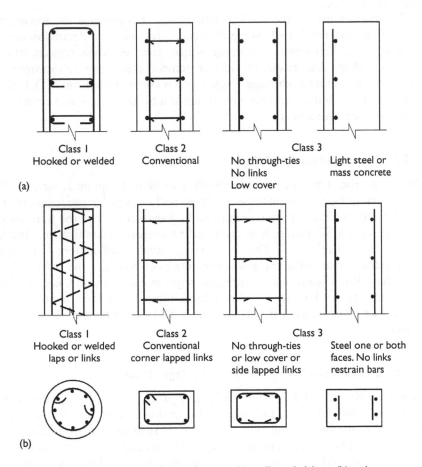

Figure 5.9 Reinforcement detailing classes – (a) walls and slabs – (b) columns

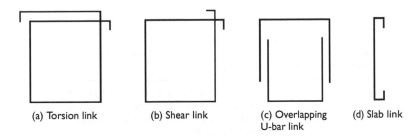

Figure 5.10 Different anchorage systems for links: in confinement terms, (a) is the most effective and (d) the least

common is type (b); the most effective is type (a) and the least effective are types (c) and (d).

The importance of the confinement effect of transverse reinforcement cannot be over-emphsised. Some elements have sensitive details owing to geometry e.g. half-joints; anchorage zones; areas of stress concentration; punching shear details; etc. In older structures, there may be examples of bent-up bars, short anchorage lengths or no conventional links. Detailing needs to be looked at carefully in arriving at a SISD rating – and even more so in detailed assessment.

5.5.5 Consequences of failure

This is a crucial factor, but one which is difficult to quantify or classify. At the preliminary assessment stage, there will already be awareness of the nature and function of the structure and its sensitivity. For the purpose of arriving at a SISD rating, it is proposed to keep the classification simple, as shown in Table 5.15. This is a very general classification, but hopefully sufficient to make the assessment engineer consider the implications of failure. What constitutes failure also requires consideration, e.g. spalling concrete may not be a serious threat for a minor bridge over a stream, but of serious consideration if it occurs on a high-rise structure in an urban setting.

5.5.6 Structural element severity ratings: SISD

Based on Sections 5.5.2–5.5.5, indicative proposals for SISD ratings are given in Table 5.16, for both free and restrained expansion. The range is between negligible (n) and very severe (A). It is likely that most structural elements will be classified either as negligible or mild; only localised details, having poor reinforcement arrangements, will be considered as candidates for B (severe) or A (very severe).

Table 5.15 Consequences of failure

Consequence of failure	Definition
Slight	The consequences of structural failure are either not serious or are localised to the extent that a serious situation is not anticipated.
Significant	If there is a risk to life and limb or a considerable risk of serious damage to property.

Table 5.16a Structured element severity ratings (SISD) a) in terms of free expansion b) in terms of restrained expansion

(a) Free expansion (mm/m)

Reinforcement detailing class	<1		1 to 2		2 to 3		3 to 4		>4	
	Slight*	Significant*	Slight*	Significant*	Slight*	Significant*	Slight*	Significant*	Slight*	Significant*
1	n	n	n	n	n	n	n	n	D	D
2	n	n	n	D	D	C	C	B	C	B
3	n	n	D	C	C	B	B	A	B	A

(b) Restrained expansion (mm/m)

	<1		1 to 2		2 to 3		3 to 4		>4	
	Slight*	Significant*	Slight*	Significant*	Slight*	Significant*	Slight*	Significant*	Slight*	Significant*
All	n	n	D	C	C	B	B	A	B	A

Reinforcement Detailing Class	Restrained Expansion (mm/m)							
	<1		1 to 2		2 to 3		>3	
	Consequences of Failure							
	slight	significant	slight	significant	slight	significant	slight	significant
All	n	n	D	C	C	B	B	A

* Consequences of failure
n = Negligible   D = Mild   C = Moderate   B = Severe   A = Very Severe

5.5.7 Local micro-climate: moisture state

No allowance for moisture state is made in Table 5.16, although this was identified in Section 5.4 as a key factor. If the concrete within the structure is not wet (i.e. RH < 90 per cent) then, in general, expansion will not take place, and the SISD rating will automatically be negligible. Only wet concretes are considered in Table 5.16.

Some care is needed here. There is evidence that the severity of cracking due to ASR can increase in the presence of a high moisture content, which might occur due to extreme exposure, leaking joints, ponding or poor drainage. Since, in some of these circumstances, there is also the possibility of the ingress of external alkalis from de-icing salts or sea water, it may then be worth considering moving the SISD rating one step to the right (from Mild to Moderate etc.).

5.5.8 Stress levels due to dead and imposed loading

If the stress levels are low, then the structural significance of ASR is less than that when the stress levels are high. This is countered to some extent by higher restraining stresses then being present, which will reduce the level of expansion. Nevertheless, this is a factor to be considered in possibly modifying the basic SISD rating. The proposal is that this be done via the recommendations of the Institution of Structural Engineers [4.25] as shown in Table 5.17.

5.5.9 Deciding on the next steps

The options available at the preliminary assessment stage are shown in Figure 5.11.

Even if confident that the structure is adequate, it may be prudent to monitor to see if expansion is continuing and, if so, to take steps to minimise exposure to water. If the preliminary assessment shows the structure to be inadequate or borderline, then the normal next step is

Table 5.17 Modifications to SISD ratings, to allow for the ratio of actual imposed load to design resistance [5.25]

Applied load/Resistance	Modification to structural element severity rating
0.0 to 0.6	Decrease rating by one (e.g. B to C)
0.6 to 1.0	No change
>1.0	Increase rating by one (e.g. C to B)

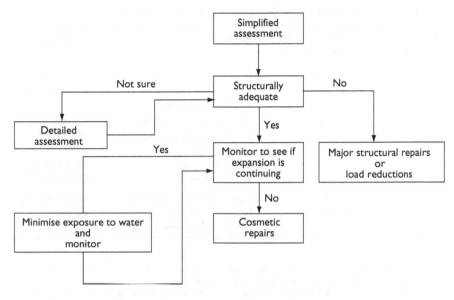

Figure 5.11 Suggested management strategy for ASR-affected structures

a detailed assessment, investigating the resistance to the various action effects, on a quantitative basis. So, how may the SISD ratings help in this decision-making process? A proposal is contained in Table 5.18, seen in relation to Figure 5.11, and applied with a measure of engineering judgement, i.e. mindful of the structural perspective dealt with in Section 5.3.

No definitive recommendations can be made for relating the SISD value to an exact residual life. This will depend on safety aspects, in comparing residual capacity with that provided in the original design and perceiving future performance requirements. Decisions to intervene may also be taken on the basis of reduced serviceability or impaired function; this is very much the province of the owner, including any decision to act for aesthetic reasons. In any assessment, even a qualitiative preliminary assessment, it is necessary to establish minimum performance requirements, which when reached will require remedial action. Having made these qualifying remarks, Figure 5.12 gives an indicative relationship between SISD rating and minimum technical performance, for ASR. It would be the author's preference to undertake a detailed assessment should the SISD be at level C or even earlier should the structure be sensitive or suffer from poor detailing. Detailed assessment is covered in Chapter 6.

Table 5.18 Relating SISD ratings (Table 5.16) to possible actions

Initial structural severity rating	Condition in the context of ASR	Action	Comments
N D	Satisfactory	Nothing beyond standard inspection routine	Easy decision
C B	Borderline	Conservative choice from: • Detailed assessment • Limited action • Monitoring • Load testing	Difficult area to decide on action, may need more investigation
A	Possibly inadequate	Remedial works and load testing depending on Detailed Assessment results	Relatively easy decision

Notes

i)   Most structures may contain elements with a combination of SISDs. Elements with similar SISDs should be grouped together for management purposes.
ii) The timescale for actions will depend on the structure, its condition, future use and life requirements. These can be broadly related to the SISD.
iii) Any strategy should be aimed at ensuring structural adequacy, "increasing life" and improving appearance. This is likely to be conservative at this stage.
iv) It may be necessary to proceed to a detailed assessment before proceeding to the stage of deciding on management options.

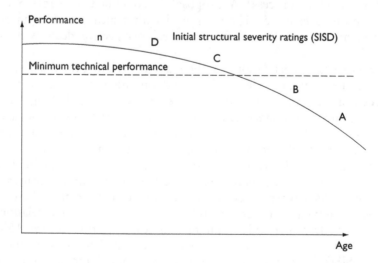

Figure 5.12 Indicative residual service lives for the range of initial structural severity (SISD) ratings

5.6 Preliminary assessment for frost action

5.6.1 Introduction

A brief description of freeze–thaw action is given in Section 4.3.7. Much more detail is given in the annexes to the relevant CONTECVET Manual [5.8], covering both the mechanism itself and the effects that it can have on structural performance. Much of this is derived from experience and research in Scandinavia, where it is treated more seriously than in the UK; this research continues, e.g. reference [5.15].

There are two types of frost damage, listed below, caused by freezing of water inside the concrete, where the water content exceeds a critical value.

Internal mechanical damage, depending on the pore structure, but generally involving micro-cracking, which reduces the mechanical properties of the concrete. When, and if, wider cracks reach the surface, the pattern is random, but with a tendency for some surface concrete to dislodge, in time; with poor quality concrete, this can lead to disintegration. Repeated freeze–thaw cycles can cause cumulative damage, spreading both horizontally and deeper into the structural element.

Surface scaling is caused by freezing of the concrete surface when in contact with saline solutions of weak concentration. The precise mechanism is not fully understood, as indicated in Section 4.3.7, but the layering of frozen and unfrozen zones is a reasonable hypothesis. The initial scaling occurs in the cement paste phase, while the aggregate grains are initially intact. As the scaling deepens, the coarser aggregate grains are lost, and in serious cases, a substantial portion of the cover can be eroded, with implications for corrosion risk and loss of bond.

The approach to preliminary assessment for frost action is based substantially on the CONTECVET Manual [5.8]; that in turn, evolved from earlier development work on ASR [4.25, 5.1], i.e. based on SISD ratings. Of the key factors identified in Section 5.4, all still obtain, but there is a difference in emphasis. For ASR, expansion was the central feature; for frost damage, micro-climate and moisture state are dominant. This leads to some different definitions compared with Section 5.5, without altering the basic approach in Section 5.4 – and with a practical need to consider internal mechanical damage and surface scaling separately.

5.6.2 Classification of consequences of failure

Here, the approach is similar as that for ASR, i.e. two classes are used: Slight and Severe. However, the definitions are a little different as shown in Table 5.19. As for ASR, variations on these definitions are perfectly acceptable in individual cases, in trying for greater precision – provided that enough detailed information is available to justify distinguishing into more than two classes.

Table 5.19 Classification of consequences of failure for frost action

Consequence of failure	Definition	Examples
Slight	Structural failure of the entire structure, or parts of it, will cause no, or small, risk of damage to property and no risk of damage to people	Pieces falling from a hydraulic structure to water or ground
Severe	Structural failure of the entire structure, or parts of it, will cause big risk of damage to people and property	Breakage of a dam causing flooding downstreams. Falling of big pieces of façade to the street

5.6.3 *Internal mechanical damage*

The major concern due to frost action is the potential reduction in mechanical properties. Even at the preliminary assessment stage, it is desirable to have some idea of the order of magnitude of this reduction. That is covered in Section 5.6.3.1. Next it is important to classify the local environment in terms of how this affects the moisture state inside the concrete; that is dealt with in Section 5.6.3.2. Finally, a structural severity rating is assigned: SISD. This is contained in Section 5.6.3.3 in very simple terms, based on the evidence of physical damage (cracking etc.) and on the criticality of the moisture state and the likely freeze–thaw regime. It is really a guide as to whether or not a detailed assessment is necessary, and, for frost action, the author would recommend that a quantitative evaluation should be made for SISD ratings of severe and very severe.

5.6.3.1 *Reduction in mechanical properties – indicative values*

Indicative values are given in Table 5.20. These are based on extensive research reported in reference [5.8]. Compressive strength is less affected than tensile strength. No values are included for elastic modulus, where the reduction can be greater still (as with ASR – Table 5.13), and for severely damaged concrete, deformation calculations are essential at the detailed assessment stage. At this point, the bond strength values are ball-park figures only; bond will be covered in greater detail in Chapter 6.

Initial strengths may be taken from the original design, should this information be available. However, as for ASR (Table 5.13), a better estimate of the undamaged concrete strength can be obtained from cores.

Table 5.20 Approximate reductions in mechanical properties due to frost action [5.8]

Strength type	1 Relation between[a] reduced and initial strength (%)	2 Biggest reduction in strength (%)	3 Lowest strength[b] (MPa)
Compressive	$(1-20/f_{c,o}) \times 100$	35	20
Split tensile	$(3-11/f_{t,o}) \times 100$	70	1
Bond strength[c] ribbed bars	$(10-35/f_{t,o}) \times 100$	70	3
Bond strength[c] plain bars	$(2.4-10/f_{t,o}) \times 100$	100	0

a $f_{c,o}$ and $f_{t,o}$ are the initial compressive strength and split tensile strength before frost damage. The relation is limited to the values in column 2.

b The lowest observed strength of severely frost damaged concrete with an initial compressive strength above 35 MPa, and an initial split tensile strength above 3 MPa.

c Bond strength is the intrinsic bond strength between a bar and concrete, no consideration given to the effect of cover and confinement by stirrups.

5.6.3.2   Classification of the environment

Moisture state in the concrete is crucial in affecting the nature, scale and extent of frost damage. In theory, for the 9 per cent expansion in the formation of ice to cause significant internal damage, the concrete should be close to saturation. In practice, damage can occur even when periods of drying take place. For that reason, a careful definition of moisture state is important. The CONTECVET proposal is shown in Table 5.21. This is based on its effects on the moisture level inside the concrete. The higher

Table 5.21 Classification of the environment for internal frost damage

Environment	Moisture characteristics	Examples
Moist	Outer: periods of exposure to water followed by longer periods of drying Inner: No accumulation of water over time	Vertical parts of facades Rain protected parts of structures e.g. slab soffits exposed to air
Very moist	Outer: long periods of exposure to water followed by periods of drying Inner: an increase in water content with time	Horizontal surfaces exposed to rain Hydraulic structures well above the water level
Extremely moist	Outer: constant exposure to water, no drying periods Inner: a significant increase in water content with time	Foundations in ground water. Bridge piers in freshwater. Hydraulic structures close to the water line

the moisture level, the greater the risk of frost damage. Low water levels, below that given for 'Moist' in Table 5.21, are unlikely to cause internal mechanical damage.

5.6.3.3 Structural element severity ratings: SISD

Indicative proposal for SISD ratings is given in Table 5.22. There is no allowance for 'reinforcement detailing class' in its role of confining expansion. Little is know about that with respect to frost damage. If frost damage does occur, then it will lead to reduced mechanical properties, with the magnitude of that dependent on environment (Table 5.21). In extreme cases, where the concrete begins to crumble and disintegrate, reinforcement detailing does become important, when bond and anchorage are seriously threatened.

The main function of Table 5.22 is to give guidance on whether or not a quantitative assessment is required, based on reduction in mechanical properties. Unlike for ASR, it is not possible to produce the equivalent of Table 5.18 and Figure 5.12, directly relating action to SISD rating. In general, a quantitative assessment should be made for S and VS ratings in Table 5.22, i.e. a move towards the detailed assessment route in Figure 5.11.

5.6.4 Salt frost scaling

Important factors in the preliminary assessment are:

– the present depth of scaling
– exposure to sea water or de-icing salts in the future
– the consequences of failure

Taking these into account, Table 5.23 gives indicative values for SISD ratings. An allowance is made for the perceived/measured depth of the scaling as a percentage of the cover, and the symbols used are the same as in Table 5.22.

Table 5.22 SISD ratings for internal frost damage

Environment	Consequences of failure	
	Slight	Severe
Moist	n	M
Very moist	M	S
Extremely moist	S	VS

n = negligible;   M = moderate;   S = severe;   VS = very severe

Table 5.23 SISD ratings for salt frost scaling

Scaling % of cover[a]	Environment	Consequences of failure[b]	
		Slight	Severe
<25	No salt	n	n
	Indirect salt spray, or sea water spray	n	M
	Direct sea water exposure	n	M
	De-icing salt exposure above −10°C	M	S
	De-icing salt exposure down to −25°C	S	VS
25–50[b]	No salt	M	S
	Indirect salt spray, or sea water spray	M	S
	Direct sea water exposure	S	VS
	De-icing salt exposure above −10°C	S	VS
	De-icing salt exposure down to −25°C	S	VS

[a] The table can only be used if the residual cover after scaling is above 20 mm. If it is lower, a quantitative preliminary assessment, or detailed assessment, shall be made.

[b] If the scaling is bigger than 50% of the cover, a quantitative preliminary assessment or a detailed assessment shall be made.

5.6.5 Deciding on the next step

For salt frost scaling, this is straight forward, since the depth of the scaling can be measured or estimated by inspection of the structure. The significance of that depth can then be evaluated with respect to:

- its influence on the strength, stiffness and serviceability of the structure owing to the reduced section area;
- its influence on the bond and anchorage of reinforcement in extreme cases;
- the increased risk of corrosion owing to the reduced cover.

For internal mechanical damage, the key issue is the magnitude of the reduced mechanical properties (Table 5.20). In Section 5.6.3.3, it is suggested that some form of quantitative evaluation should be carried out for SISD ratings of S and VS (Table 5.22). Initially, this may simply be an attempt to get a better estimate of the reductions than that provided by Table 5.20, which gives lower bound values. This alone may give reassurance. However, in cases of more severe damage, it may be necessary to investigate individual action effects such as bending, shear, compression and bond; the decision to do so will also depend on structural sensitivity. Detailed assessment is covered in Chapter 6.

Table 5.24 Some examples of synergy, due to deterioration mechanisms acting simultaneously

Combination of mechanisms	Possible effects
Surface scaling due to frost and corrosion	This may lead to a gradual reduction of the cover to the reinforcement and, hence, increases the likelihood of corrosion
Alkali–silica reaction, and either frost action or corrosion	The expansive action of ASR may lead to wide cracks which can fill with water, and which, if frozen, may cause internal mechanical damage. This same action may also permit easier access to the reinforcement of water containing chlorides, causing more severe corrosion. On the other hand, gel caused by ASR may fill pores, thus densifying the cement matrix
Leaching and frost action	The influx of water may increase the moisture uptake and, hence, reduce the internal frost resistance
Leaching and corrosion	The leaching of lime from the concrete cover increases the rate of carbonation and the diffusivity of chlorides and reduces the critical threshold level

5.6.6 Synergetic effects

If several destruction mechanisms are acting simultaneously with frost, then the recommendations given here have to be applied carefully. In hydraulic structures, for example, if leaching is also occurring, then the Table 5.20 reductions can be greater. Some other examples of synergetic effects are given in Table 5.24. Assessment is always based on establishing what is the dominant deterioration mechanism and following the procedures for that. Occasions may arise, however, where the interplay between mechanisms is significant, and this will require a separate assessment.

5.7 Preliminary assessment for corrosion

5.7.1 Introduction

In assessment for the effects of corrosion, compared with ASR or frost, there is a greater need to get to know the structure and its various sensitivities. Corrosion directly affects the reinforcement, which may be acting either in tension or compression. This reinforcement bonded to and anchored in the concrete, will have been designed to control the strength, stiffness and serviceability of structural elements. It follows that any deterioration may be more significant, and the nature of the corrosion mechanism inevitably means a different emphasis in assessment – while still following the basic principles in Section 5.4 for preliminary assessment.

Earlier sections in this book (3.2.4.2; 4.3.4) touched on corrosion without pretending to be comprehensive regarding the nature of the mechanisms; reliance is put on the references for that level of detail. The focus is on structural assessment, beginning in Section 4.6, continuing in Section 5.3.5, and now moving on to preliminary assessment involving the SISD ratings approach in the CONTECVET manuals [5.1, 5.8, 5.13].

5.7.2 Consequences of failure

Again, this is a central feature in deriving SISD ratings. For corrosion, the proposal is to adopt the same definitions for "Slight" and "Significant", as given in Table 5.15 for ASR, with the notes in Section 5.5.5 also applying.

5.7.3 Basis for SISD ratings for corrosion

For ASR and frost action, there was one dominant factor from the list given in Section 5.4, which controlled the make up of the SISD ratings: expansion for ASR; moisture state for frost action. The situation is not so clear-cut for corrosion, and, in the CONTECVET manual [5.13], it was proposed that the SISD rating should be derived from two sub-components as follows:

(1) A simplified corrosion index (SCI), which took account of both the aggressivity of the environment and the observed/measured deterioration, either in material or structural terms. In this way, some indication can be obtained of both the current reduction in mechanical and section properties and the potential future rate of deterioration.
(2) A simplified structural index (SSI) which took account of structural sensitivity to the corrosion process. This is analogous to the structural detailing factor for ASR as given in Section 5.5.4.

Thus:

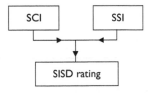

The derivation of these sub-components, and their integration into SISD ratings is covered in Sections 5.7.4–5.7.6.

5.7.4 Simplified corrosion index (SCI)

As indicated in Section 5.7.3 [1] above, this has two components:

1. a corrosion damage index (CDI), and
2. an environmental aggressivity factor (EAF).

5.7.4.1 Corrosion damage index (CDI)

There are various factors which could be included to evaluate this index, depending on the information available from inspection and in situ testing. To illustrate the process, six factors have been selected as follows:

 (i) the depth of penetration of the aggressive front (chlorides or carbonation);
 (ii) the concrete cover;
(iii) extent of cracking due to corrosion;
(iv) the presence of rust and any observed loss of rebar area;
 (v) measurement/estimation of the corrosion rate; and
(vi) concrete resistivity.

These are given weights 1–4 respectively for the four levels shown in Table 5.25.

Table 5.25 Suggested corrosion indicators and weightings for the four proposed levels of damage, in deriving CDI values

Damage Indicators	Level I (1)	Level II (2)	Level III (3)	Level IV (4)
Carbonation depth	$X_{CO2} = 0$	$X_{CO2} < c$	$X_{CO2} = c$	$X_{CO2} > c$
Chloride level	$X_{Cl}^- = 0$	$X_{Cl}^- < c$	$X_{Cl}^- = c$	$X_{Cl}^- > c$
Cracking due to corrosion	No cracking	Cracks w < 0.3 mm	Cracks w > 0.3 mm	Spalling and generalised cracking
Resistivity (Ω.m)	>1000	500–1000	100–500	<100
Bar section loss	<1%	1–5%	5–10%	>10%
Corrosion rate of main reinforcement (μA/cm^2)	<0.1	0.1–0.5	0.5–1	>1
Weighting	1	2	3	4

Notes

- X_{CO2} is the actual carbonation front in [m].
- X_{Cl} is the actual chloride threshold front in [m].
- C is the concrete cover in [m].
- W is the crack width in [mm].

5.7.4.2 Environmental aggressivity factor (EAF)

This is more difficult to evaluate, since the local micro-climate and moisture state can vary considerably on a seasonal basis. Information on water permeability, surface chloride levels, half-cell potential, resistivity etc. may be available, and, if so, should be used. If not, a fall-back position is the definition of exposure classes given in design Codes and Standards. These are now quite detailed and Table 5.26 gives an abbreviated version of the classifications used in EN 1992–1 [5.16] and BS 8500 [5.17].

On the right hand side of Table 5.26, values are suggested for weighting factors to represent the aggressivity of the different classes on a scale 0–4.

5.7.4.3 Evaluating the SCI from Tables 5.25 and 5.26

This essentially empirical approach is best illustrated by an example. As will be seen later, the objective is to come up with an SCI value between 0 and 4. Let us suppose that all the factors in Table 5.25 are assessed as being level 2, giving an average weighting factor of 2 for CDI. From Table 5.26, let us suppose that the most realistic representation of the actual environmental conditions is XD3, carrying a weighting factor of 4 for environmental aggressivity (EAF). It is then proposed that these be averaged to obtain a value for the SCI. Thus

$$SCI = \frac{CDI + EAF}{2} = 3 \text{ for this particular example}$$

This may appear overly simplistic, but it is possible to obtain a perspective by choosing weightings for the six factors in Table 5.25 from different Levels and to mix these with different exposure classes from Table 5.26. The objective is to provide some guidance on whether or not a detailed assessment is necessary, as covered in Chapter 6, and on the urgency of intervention, if any.

5.7.5 Simplified Structural Index (SSI)

Table 5.14 and Figures 5.9 and 5.10 gave brief details of different reinforcement classes in terms of their ability to contain the expansive ASR. Considerable emphasis was put on the nature and layout of the transverse reinforcement and on the effectiveness of its anchorage. Similarly, with corrosion, there is a need to take account of the efficiency of the detailing in deriving SISD ratings.

Table 5.26 Classification of exposure classes taken from BS 8500 [5.17] with suggested weighting factors for assessing the environmental aggressivity factor (EAF) (Permission to reproduce extracts from the British Standards is granted by BSI)

Class designation	Class description	Informative examples applicable in the United Kingdom	Environment aggressivity factor (EAF)
No risk of corrosion or attack (X0 class)			
X0	For concrete without reinforcement or embedded metal: all exposures except where there is freeze-thaw, abrasion or chemical attack	Unreinforced concrete surfaces inside structures	
		Unreinforced concrete completely buried in non-aggressive soil	
		Unreinforced concrete permanently submerged in non-aggressive water	
		Unreinforced concrete in cyclic wet and dry conditions not subject to abrasion, freezing or chemical attack	0
	For concrete with reinforcement or embedded metal: very dry	Reinforced concrete in very dry conditions	
Corrosion induced by carbonation (XC classes)			
(where concrete containing reinforcement or other embedded metal is exposed to air and moisture)			
XC1	Dry or permanently wet	Reinforced and prestressed concrete surfaces inside structures except areas of structures with high humidity. Reinforced and prestressed concrete surfaces permanently submerged in non-aggressive water	1
XC2	Wet, rarely dry	Reinforced and prestressed concrete completely buried in non-aggressive soil	2
XC3 and XC4	Moderate humidity or cyclic wet and dry	External reinforced and prestressed concrete surfaces sheltered from or exposed to, direct rain	2 (XC3)
		Reinforced and prestressed concrete surfaces inside structures with high humidity (e.g. bathrooms, kitchens)	
		Reinforced and prestressed concrete surfaces exposed to alternate wetting and drying	3 (XC4)

Corrosion induced by chlorides other than from sea water (XD classes)

(where concrete containing reinforcement or other embedded metal is subject to contact with water containing chlorides, including de-icing salts, from sources other than sea water)

XD1	Moderate humidity	Concrete surfaces exposed to airborne chlorides	2
		Reinforced and prestressed concrete surfaces in parts of bridges away from direct spray containing de-icing agents
		Parts of structures exposed to occasional or slight chloride conditions
XD2	Wet, rarely dry	Reinforced and prestressed concrete surfaces totally immersed in water containing chlorides	3
XD3	Cyclic wet and dry	Reinforced and prestressed concrete surfaces directly affected by de-icing salts or spray containing de-icing salts (e.g. walls, abutments and columns within 10 m of the carriageway, parapet edge beams and buried structures less than 1 m below carriageway level, pavements and car park slabs)	4

Corrosion induced by chlorides from sea water (XS classes)

(where concrete containing reinforcement or other embedded metal is subject to contact with chlorides from sea water or air carrying salt originating from sea water)

XS1	Exposed to airborne salt but not in direct contact with sea water	External reinforced and prestressed concrete surfaces in coastal areas	2
XS2	Permanently submerged	Reinforced and prestressed concrete completely submerged and remaining saturated, e.g. concrete below mid-tide level	3
XS3	Tidal, splash and spray zones	Reinforced and prestressed concrete surfaces in the upper tidal zones and the splash and spray zones	4

While Figures 5.9 and 5.10 might provide a useful starting point, the needs are rather different in relation to corrosion. Should the observed/measured deterioration be severe (say levels III and IV in Table 5.25), then the presence of well anchored links is crucial in maintaining bond and anchorage; there is a wealth of experimental evidence for this in the literature. Further, should severe spalling or delamination occur, the need to contain the residual concrete section is also important. Finally, if the confinement creates a tri-axial stress situation, thus increasing residual strength and providing greater rotation capacity, while possibly creating alternative load-bearing mechanisms, then anchorage of the transverse steel becomes doubly important.

To evaluate an element's ability to cope with the above issues, it is proposed that Figures 5.9 and 5.10 be 'converted' into numerical values that reflect their ability to act effectively when corrosion damage occurs. To do this simply is not easy, since there are many qualifying features. The approach here is to provide a table of base values and add some qualifications and footnotes. This is done in Table 5.27.

Reinforcement details can vary a great deal, either as originally conceived or as built. This is not an exact science, but an attempt to gain insight in deciding what the next step should be. In the interests of simplicity, the indices have been kept in the range 0–4, in line with Tables 5.25 and 5.26. Some modifications have been proposed to the basic SSI values, e.g. where small links are used or where the cover to main column bars is substantial (greater risk due to spalling and to buckling of the main bars). The real use of Table 5.27 is in combining it with SCI values (Section 5.7.4) to produce initial SISD ratings.

5.7.6   Structural element severity ratings (SISD) for corrosion

Based on inputs from Sections 5.7.4 and 5.7.5, Table 5.28 suggests some values for SISD ratings, expressed in terms of negligible, medium, severe and very severe. In compiling this table, the author has reverted to the reinforcement detailing classes in Figure 5.9, rather than using the numerical SSI values in Table 5.27. While the SSI values in Table 5.27 give a better approximation of anchorage efficiency in meeting the performance needs in Section 5.7.5, it is difficult to make this level of distinction in deriving SISD ratings in terms of negligible, medium severe and very severe.

It is suggested that for all ratings of severe and very severe, a detailed assessment be carried out. If there is a reason to believe that the corrosion

Table 5.27 Suggested Simplified Structure Indices for different arrangements of transverse reinforcement

	No links or through ties	Compliance with Code requirements for link spacing assumed			Torsion links; spiral reinforcement (class I in Figure 5.9)
		Over lapping u-bars (class 3 in Figure 5.9)		Conventional stirrups and links (class 2 in Figure 5.9)	
		Without hooks	With hooks		
Basic SSI values (beams and columns)	0	1	2	3	4
Ratio of link dia . to main steel ≤ 0.4	–	0	1	2/3	3/4
Columns: ratio of core concrete to overall section area ≤ 0.75	–	0	1	2/3	3/4

Notes
1 Anchorage of transverse reinforcement is paramount.
2 For columns, if link spacing is greater than current design requirements, use the lower values in columns 5 and 6.
3 Consideration should be given to the ratio of actual imposed load to theoretical ultimate strength.
4 Particular care is necessary for special elements such as corbels, or in cases involving punching shear.

Table 5.28 Suggested values for SISD ratings for corrosion

SCI value	Simplified Structural Index (SSI) (assuming compliance with Code rules for link spacing)							
	Class 1		Class 2		Class 3		No links	
	Consequences of failure							
	Slight	**Significant**	**Slight**	**Significant**	**Slight**	**Significant**	**Slight**	**Significant**
0–1	n	n	n	n	m	m	m	m
1–2	n	m	n	m	m	m	m	S
2–3	m	m	m	S	S	S	S	VS
3–4	m	S	S	S	VS	VS	VS	VS

n = negligible   S = Severe
m = medium   VS = Very severe

rate is high, it may also be necessary for medium ratings, particularly for sensitive structures. Tables 5.25 and 5.26 should also be taken into account when the damage indicators are mostly level 3 or level 4, and the exposure class is severe.

5.8   The nature and timing of intervention

Intervention, which is the subject of Chapter 7, comes generally after a Detailed Assessment (Chapter 6) that clearly demonstrates that positive action is necessary. However, some brief remarks are relevant at this level 3 stage (Figure 5.1), since the options available (and their timing) are many and varied, within the full spectrum of doing nothing to knocking down the structure.

What action is taken at this stage in an assessment will depend on the maintenance regime, management strategy and future plans for the structure. Various possibilities appear at different stages in the Figure 5.1 flow diagram. Monitoring is a useful technique, plus the introduction of preventative measures at a relatively early stage to slow down or even stop the deterioration process. Protective coatings of various kinds come into this category to control the moisture state in the concrete, as do techniques such as cathodic protection or chloride extraction. An ideal mix of minimum maintenance cost and technical efficiency is the overall objective.

In pure structural terms, the consequences of failure will always loom large. Much will depend on what factors in Table 5.6 are considered to be especially at risk, while bearing in mind Tables 5.8 and 5.9. A preliminary risk assessment of the type illustrated in Figure 5.10 may be helpful here, in support of the qualitative SISD approach for the dominant deterioration mechanism.

The timing of any intervention is inevitably subjective, depending on what is acceptable to the owner in terms of minimum technical performance. Provided that safety levels are satisfactory, some may wish to act early in preventative mode, others may be prepared to wait, perhaps while introducing monitoring into the normal inspection and maintenance regimes. There is no single universal answer to this, but, in the chapters that follow, an attempt will be made to give guidance on establishing minimum performance criteria and on indicative rules for decision-making.

References

5.1 CONTECVET IN30902I. A validated Users Manual for assessing the residual service life of concrete structures affected by ASR. A deliverable from an EC Innovation project. Available from the British Cement Association (BCA), Camberley, UK. 2000.

5.2 The Concrete Society. *Diagnosis of deterioration in concrete structures*. Technical Report 54. 2000. The Concrete Society, Camberley, UK.

5.3 Concrete Bridge Development Group (CBDG). *Guide to testing and monitoring the durability of concrete structures*. Technical Guide No. 2. 2002. CBDG, Camberley, UK.

5.4 Bungey J.H. and Millard S.G. *Testing of concrete in structures*. Blackie Academic and Professional, Glasgow, UK. 3rd Edition 1995. p. 286.

5.5 British Cement Association (BCA). *The diagnosis of alkali-silica reaction*. Report of a Working Party. Report No. 45.02 (2nd Edition). 1992. BCA, Camberley, UK.

5.6 Clark L.A. *Critical review of the structural implications of the alkali-silica reaction*. Transport Research Laboratory (TRL). Contractor Report 169. 1989. TRL, Crowthorne, UK.

5.7 Institution of Structural Engineers. *Structural effects of alkali-silica reaction*. Technical guidance on the appraisal of existing structures (2nd Edition). IStructE, London, UK. 1992.

5.8 CONTECVET IN30902I. A validated Users Manual for assessing the residual service life of concrete structures affected by frost action. A deliverable from an EC Innovation project. Available from the British Cement Association, Camberley, UK. 2000.

5.9 The Concrete Society. *Non-structural cracks in concrete*. Technical Report 22. December 1982. The Concrete Society, Camberley, UK.

5.10 The Concrete Society. *The relevance of cracking in concrete to corrosion of reinforcement*. Technical Report 44. 1995. The Concrete Society, Camberley, UK.

5.11 Webster M.P. The assessment of corrosion-damaged concrete structures. PhD Thesis. University of Birmingham, UK. July 2000.

5.12 Concrete Bridge Development Group (CBDG). *Notes for guidance on the assessment of concrete bridges*. Technical Guide No. 9. 2006. CBDG, Camberley, UK.

5.13 CONTECVET IN30902I. A validated Users Manual for assessing the residual service life of concrete structures affected by corrosion. A deliverable from an EC Innovation project. Available from the British Cement Association, Camberley, UK. 2000.

5.14 Jones A.E.K. and Clark L.A. The practicalities and theory of using crack width summation to estimate ASR expansion. *Proceedings of the Institution of Civil Engineers; Structures and Buildings*. Vol. 104. 1994. pp. 183–192. ICE, London, UK.

5.15 Fridh K. *Internal frost damage in concrete : experimental studies of destruction mechanisms.* Report TVBM – 1023. 2005. Lund Institute of Technology, Department of Building Materials. Lund, Sweden.

5.16 British Standards Institution (BSI) BS EN 1992–1–1. Eurocode 2: design of concrete structures – Part 1: general rules and rules for buildings. BSI, London, UK. 2004.

5.17 British Standards Institution (BSI). Concrete – complementary British Standard to BS EN 206–1. BS 8500. BSI, London, UK. 2002.

Chapter 6

Detailed structural assessment

6.1 Introduction

If the nature/extend of the damage, linked to qualitative SISD ratings, has indicated that further investigation is necessary before deciding on the nature and timing of any intervention, then the logical next step is a detailed assessment. This involves a more quantitative approach, with the emphasis on structural performance – strength, stiffness, serviceability, functionality – and on how this might be affected by deterioration. Before rushing into detailed calculations, it is prudent to recap on the overall perspective given in Section 5.3. There, the focus was on:

(a) Performance requirements to be looked at – an appropriate selection from Table 5.6, under the three major headings.
(b) Structural sensitivity – issues such as those in Tables 5.7 and 5.8, while being mindful of the design basis, the consequences of failure and the quality of construction and maintenance regimes.
(c) The nature and sophistication of the analytical method(s) to be used (Section 5.3.3), while also estimating the actual imposed loads (compared with those assumed in design). A critical feature here is the selection of section and mechanical properties as input, which represent the structure as built and as affected by the deterioration.
(d) The material and structural parameters likely to be affected by the dominant deterioration mechanism (Tables 5.9–5.12), i.e. moving towards a preliminary risk assessment.
(e) A perspective of the particular case of corrosion damage (Section 5.3.5). Corrosion is unique among deterioration mechanisms because of its direct effect on the reinforcement and prestressing steel.

Items (a)–(e) above are all important in terms of providing a perspective, but not the whole picture. In addition, it is necessary to consider:

(f) Acceptable minimum technical performance. There will always be a strong emphasis on safety aspects, but serviceability and function (Table 5.6) may often dictate the nature and timing of any intervention.

(g) The favoured management and maintenance strategy. Some owners may prefer early intervention, mainly involving preventative/delaying measures and some making good. Others may prefer to wait, while continuing to monitor the situation, before embarking on a major rehabilitation programme. Future plans for the structure will also come into the equation, and the assessment will then require more effort in predicting likely future rates of deterioration.

Items (f) and (g) are clearly related to each other, but also closely linked to items (a)–(e). It is the overall picture which matters, and governs what is done in a detailed assessment.

Against that general background, the approach in this book will be to first look at the effects of deterioration on the stiffness and stability of the structure, and in particular on the impact of that in possibly changing the distribution of action effects (bending, shear, etc.) compared with that assumed in design, i.e. to focus on the analysis of the structure as a whole. Following on from that is the coverage of residual capacities (analysis of critical sections). For this, the influence of reductions in the mechanical and section properties of the concrete (from whatever cause) will be dealt with as a single entity.

There are two main reasons for the above approach:

1) the philosophy of keeping the approach as simple as possible, consistent with the quality of available input information; and
2) the belief that most owners will want to make direct comparisons between what was provided in the original design and what is the assessed current residual strength.

Much of the literature on assessment is based on theoretical methods involving risk analysis, probabilistic methods and reliability indices. The author's approach is essentially deterministic, but, in using modified design models, there is an underlying link to reliability indices, as clearly indicated in EN 1990 [6.1]. The partial factor approach to design is derived from reliability indices found to be acceptable to both the profession and to society in general. In the interests of simplicity, it seems sensible to work directly with modified design models in the first instance, while keeping the option of modifying reliability indices in reserve – as done by the Canadian Bridge Code, for example (Section 3.2.1.3; reference [3.11]).

It is realised that modern sophisticated methods of analysis, such as those involving non-linear finite elements, though benefiting from re-distribution, merge the analysis for load effects with capacity calculations. These methods are perceived as follow-up approaches, aimed at seeking improvements in 'assessment ratings', and will be touched on briefly later in this chapter.

The scope and contents of this chapter is as follows:

- 6.2 Physical effects of deterioration
- 6.3 Minimum technical performance
- 6.4 The use of mechanical properties of materials in design
- 6.5 Analysis of structures
- 6.6 Overview of the effects of deterioration on the strength of elements and sections

6.2   Physical effects of deterioration

Table 5.6 provides a list of performance requirements that may need checking. What then are the key effects of deterioration influencing these? Table 6.1 provides a brief summary, together with some examples and comments.

Detailed assessment is concerned primarily with the adequacy of structural performance and how to evaluate any reductions in the performance of sections, elements or the structure as a whole. Some of the items listed in Table 6.1 may require remedial action in any case, for reasons of function, appearance or serviceability in general.

Before moving onto the evaluation of reductions in performance requirements in later sections of this chapter, it is worth looking at a simple example of the proposed approach. Consider the rectangular section shown in Figure 6.1, with the normal design assumptions for maximum strain in the concrete in compression and a possible compressive strength block, acting in flexure at the ultimate limit state.

For the undamaged condition, the design resistance is given either by:

$$M = A_s f_{yd}(d - \beta x), \text{ where the reinforcement controls} \tag{6.1}$$

or

$$M = f_{av} b x (d - \beta x), \text{ where compression controls} \tag{6.2}$$

For a damaged condition, then, any or all of the following might be affected:

$$x, b, d, f_{av}, A_s, f_{yd}, \beta, 0.0035 \tag{6.3}$$

The objective is to determine which, and by how much, and then to modify the appropriate equation for undamaged condition accordingly.

Table 6.1 Examples of physical effects of deterioration

Category	Examples	Comments
1. Loss of section	Surface scaling due to frost action	In extreme cases, can lead to the onset of corrosion.
	Spalling due to expansive corrosion products or to ASR	In extreme cases, can lead to delamination, and to a major influence on bond and anchorage, leading to reduced strength and/or alternative load-carrying mechanisms.
	Reinforcement – reductions in cross-sectional area, due to corrosion	A major concern, together with the spalling issues above.
2. Reductions in mechanical properties	Concrete strength	Important for evaluating critical sections, where the effect may be different for different action effects. Virtually all deterioration mechanisms, which involve expansion, reduce strength (Tables 5.9 and 5.10).
	Element and structure stiffness	Important for: – analysis of structure (distribution of load-effects) – assessing serviceability factors (Table 5.6) Reductions in elastic modulus tend to be greater than for strength.
	Ductility	There is some evidence that corrosion can affect the reinforcement, in limiting elongation and the ratio between yield and ultimate strength. This is important when re-distribution is required.
3. Excessive deformation	The creation of alternative modes of failure in extreme cases	Figure 5.3 is a particular example. Also, the possibility of cover spalling in columns, changing the behaviour mode from that for a short column to a slender one. Serviceability may also be a factor, in terms of cracking, e.g. for appearance or liquid retention.
	Overstress	In the presence of restraints in heavily loaded structures, expansion can lead to overstress.
	Local detailing failures	At areas of stress concentration such as bearings or re-entrant angles, excessive deformation and/or loss of strength can lead to local failures. Here, it is also prudent to look carefully at the real performance of expansion joints and the articulation system in general.

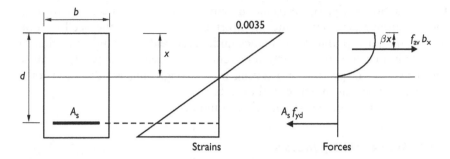

Figure 6.1 Typical design assumptions for a rectangular section in bending at the ultimate limit state

This is the simplest possible example, although a similar approach would be used for columns. Actions such as shear and bond can be more complex for two main reasons:

(1) The design equations are empirical and may not reflect real performance in the damaged condition; other models may be more appropriate.
(2) The damage is likely to affect some elements in the design equations differently, compared with flexure; e.g. concrete tensile strength will be affected differently compared with compressive strength. Moreover, the influence of stirrups on bond strength is virtually impossible to quantify analytically, and recourse has to be made to experimental evidence.

Having made these qualifications, Figure 6.1 still represents the principles of the approach to be used later. This is a two-stage operation:

(1) from inspections and testing, to derive as much information as possible on the nature and extent of the physical damage (Table 6.1).
(2) based on this information, to know how to modify the design equations.

6.3 Minimum technical performance

6.3.1 Introduction

Figure 5.1 recognises the need to establish what constitutes minimum acceptable technical performance, before moving on from a preliminary assessment to a detailed one, and figures in earlier chapters of this book illustrate different aspects of that, e.g. Figures 3.3, 3.13, 4.8 and 5.12.

Some owners may wish to do this earlier, say at levels 1 or 2 in Figure 5.1, either because the visual damage is unacceptable for aesthetic and functional reasons (Table 5.6) or because there is the possibility of more serious deterioration occurring relatively quickly, and the management strategy is to take action early, by introducing some preventative measures to augment routine maintenance and monitoring. This suggests the need to look at minimum acceptable performance under serviceability conditions first of all, partly because of the early interventions strategy, and partly because the data from inspection and testing may strongly suggest the need for a detailed assessment on structural aspects.

This section is presented with that in mind.

6.3.2 Serviceability factors

There are three separate considerations here:

(i) reductions in section or mechanical properties which affect the stiffness of the structure, leading to a different distribution of action effects to that assumed in design, and possibly to excessive deflections or deformation;

(ii) the whole question of cracking from whatever cause (Figure 4.3); and

(iii) the particular case of corrosion risk where inspection, testing and diagnosis suggests that the condition is still at level 1 in Table 5.25, but that higher levels are imminent.

6.3.2.1 Reductions in section or mechanical properties

Reductions in section properties are usually fairly obvious and visible; surface scaling due to frost action is a particular example, so are the effects of abrasion, erosion or weathering. At a low level of damage, a decision to take action for this reason alone is largely subjective. At a higher level, in combination with reduced mechanical properties, it becomes a more serious structural issue, and will be dealt with in a later section of this chapter.

Reductions in mechanical properties are more difficult to evaluate. It will be common to take cores for testing, particularly if an expansive action such as ASR or frost is suspected. These will give a measure of compressive strength primarily, but can also indicate values for elastic modulus and tensile strength. The indicative values in Tables 5.13 and 5.20 suggest that the latter two may be affected more than compressive strength, for any given level of attack. Chemical attack, such as that due to sulfates, occurs mainly in foundations (Table 4.4), and the effects are summarised in Section 4.3.5. Again, reductions in both section and mechanical properties can occur, with a tendency for the concrete to soften; probably the most serious case is thaumasite attack (Section 4.3.5.2), with the concrete turning to mulch in extreme cases.

Plainly, chemical attack can occur under service conditions, and, if suspected, the investigation would have to go beyond simply quantifying reduced mechanical properties in serious cases. The structural implications will depend on the type of structure and its sensitivity to the effects of deterioration.

The general picture that is being painted here is one of progressive investigation, via testing. The prime concern will be to establish reduction in stiffness, but if this indicates major deficiencies an immediate move towards a detailed structural assessment will follow.

6.3.2.2 Cracking

Causes of cracking are identified in Figure 4.3, with possible locations shown in Figure 4.4. Table 5.2 suggests a system for classifying cracks in terms of width and spacing, which is general, while Table 5.1 gives a damage rating approach based on crack width, focused mainly on corrosion; this is taken further in Table 5.25, in association with other damage indicators for corrosion.

In decision-making, it is not just a question of crack width and spacing. The cause of the cracking has to be identified, as well as its extent and location. As a simple guide, damage ratings of 4 or 5 in Table 5.1 warrant further investigation and possibly a detailed assessment; this is also true of levels III and IV in Table 5.25. However, the focus under serviceability conditions is on whether or not action is necessary to fill the cracks for aesthetic or functional reasons and is inevitably subjective and related to normal maintenance regimes.

6.3.2.3 Corrosion

The focus here is on situations where preliminary investigations indicate that the condition is no worse than level I in Table 5.25, i.e. corrosion has not started, but either carbonation or corrosion fronts are advancing towards the reinforcement, and the owner may wish to take remedial action at an early stage. This possibility is dealt with in Section 4.6, while indicating how Figure 4.8 might be used for different limiting-performance criteria on the vertical axis.

To take advantage of this – and it could be very beneficial, since the time to move between the different suggested criteria can be substantial – the corrosion mechanisms would have to be modelled, using measured values for the parameters included in Group C in Section 4.6, with the ability to predict the time needed to move from one criterion to the next.

6.3.3 Safety levels

Safety will always be a primary concern in assessment, and the tendency will be towards conservatism, particularly when the consequences of failure are highly significant. It is suggested that the relationship between residual capacity and that provided in the original design will always be of interest to owners, and therefore this section looks briefly at the evolution of safety levels and how the basis for these has changed.

This is best done via Codes of Practice. Over the last century, design practice has evolved from elastic methods via ultimate strength design to the limit-state approach now embodied in the Eurocodes, e.g. reference [6.2]. Design models have become more refined, and new ones introduced. Materials have become stronger and stiffer, and lower factors of safety have been used, as confidence in the technology increased. It is prudent, therefore, when assessing old concrete structures, to ascertain what standards applied at the time of construction and how these compare with current methods. There is a strong empirical element to this evolution, involving a great deal of calibration against experimental data; the elasto-plastic behaviour of concrete up to ultimate can be difficult to model, and past models do not always have a true scientific basis. This historical evolution has nevertheless produced satisfactory designs, in balancing economy against acceptable performance.

It is only relatively recently that this practical approach has been seen alongside reliability theory, in an attempt to develop Codes in a more controlled manner. The principles have been set down in ISO 2394 [6.3], the basis for the approach in BS EN 1990: 2002 [6.1], Figure 6.2, from reference [6.1], shows the relationship between the historical deterministic methods

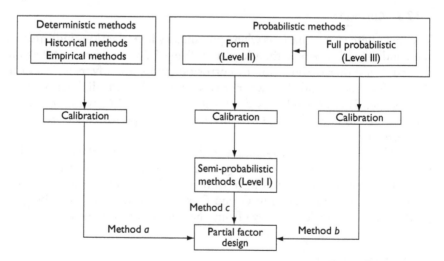

Figure 6.2 Overview of the relationship between deterministic design methods and the probability approach (Permission to reproduce extracts from the British Standards is granted by BSI [6.1])

and the probabilistic approach. Both tend towards partial factor design, with method 'a' being the norm in operational Codes and probabilistic methods being used to further develop safety aspects.

Conventionally, reliability methods use reliability indices, which are related to the probability of failure. In putting forward recommendations for partial factors, BS EN 1990 uses value for the reliability index of $\beta = 3.8$ for the majority of structures, over a 50-year reference period, and the general consensus is that this is achieved for the current set of operational concrete design Codes.

However, it is permitted to vary the reliability index; indeed, a higher value ($\beta = 4.3$) is suggested for cases where the consequences of failure are very high, and a lower one ($\beta = 3.3$) where the consequences of failure are small or negligible.

This preamble on reliability serves as a lead-in to a consideration of what might constitute minimum acceptable performance, in terms of safety, for assessment. In passing, it is interesting to note the suggested β values in Table 3.2, as recommended by the Canadian Standards Association for bridges [3.11] – and the basis for the variations; the highest values is 3.75 and the lowest is 2.25.

There are some fundamental differences between design and assessment, and Table 6.2 summarises these. In assessment, there are fewer assumptions, and it is possible to determine most of the key inputs either by measurement or by definition. As suggested in the table, this leads to the possibility of lower values for partial safety factors, without significantly altering the overall reliability.

To understand how this might be done in terms of partial safety factors, it is necessary to consider what is covered by the partial factors. In over-simplified terms, we have:

Table 6.2 Significant differences between Design and Assessment

Item	Design	Assessment
(1) Material properties	Assumed	Measured
(2) Dead loads	Calculated	Accurately determined
(3) Live loads	Assumed	Assessed
(4) Analysis	Code based	More rigorous alternatives
(5) Load effects	Bending, shear, Compression, cracking dominate	Anchorage, bond detailing may be more important
(6) Environment	Assumed classification	Definition of macro- and micro-climate
(7) Reliability	Code values for safety factors	Smaller factors (?) for same reliability

γ_F for loads

    (i) γ_{F1} uncertainty of representative values for loads
   (ii) γ_{F2} uncertainty in the models used to determine the effects of loads (bending, shear, etc.)

γ_M for materials

  (iii) γ_{M1} uncertainty in material properties
  (iv) γ_{M2} uncertainty in structural resistance

If this is related to Table 6.2, we see that:

- Items (2) and (3) relate to γ_{F1}
- Item (4) relates to γ_{F2}
- Item (1) relate to γ_{M1}
- Item (5) relates to γ_{M2} with the proviso that different load effects may be more critical in assessment, these generally having more empirical models than those for the main design load effects.

There is also potential benefit in item (5), in exploring the micro-climate in some detail for detailed assessment, although Table 5.26 contains recommendations for current environmental classifications.

What is being suggested here is the possibility of using the basics of reliability in selecting/calculating values for partial safety factors, applied loads and residual capacities, in settling on acceptable minimum safety levels. What is achievable will depend on what is measured and the precision of these measurements. In particular, there is benefit in measuring material properties (γ_{M1}) and, especially, determining better representative values for imposed loads (γ_{F1}), possibly leading to reduced values for the partial safety factor (in design, we already do this, having different values for dead and imposed loads). Modifying design resistance is more difficult, because of uncertainty in modelling the precise effects of deterioration on the various action effects.

6.4   The use of mechanical properties of materials in design

6.4.1   Introduction

The overall approach to detailed assessment in this book is to use design models modified to take account of deterioration. The testing and diagnosis phases in the investigation will involve attempts to measure reductions in section and mechanical properties. To do that properly, and in interpreting the data, it is necessary to have some appreciation of what properties are used, how they are used, and for what purpose.

6.4.2 Mechanical properties for concrete

In analysing structures to determine maximum values for action effects, the normal practice is to use elastic methods and to assume that the elements are uncracked. In calculating stiffness, the whole cross-section is used, and a value is needed for the elastic modulus, associated with the planned design strength of the concrete.

In calculating the resistance of critical sections, the basis is the characteristic strength of the concrete at 28 days, determined from standard control tests on either cylinders or cubes. Other necessary mechanical properties are then derived from assumed relationships to the characteristic strength.

Current practice on this is summarised in Table 6.3, based on BS EN 1992-1-1 [6.2]. This has the characteristic 28-day cylinder strength as a base, with relationships given for characteristic cube strength and for mean strength, the latter being a likely target in concrete mix design. The table also gives values for different assumed levels of tensile strength, with the footnotes indicating how these can be related to either flexural or splitting tensile strength. Finally, an expression is given for estimating elastic modulus, with note (5) indicating how this might be modified for different types of aggregate. Notes (2) and (4) show how design values are obtained for compressive and tensile strengths respectively.

In a sense, the characteristic strength is fictitious, coming from control specimens tested in a standard way. The actual strength in the structure will be different, depending on the type of element and the nature of the loading. Designers know this and modify their design models accordingly to provide the accepted levels of safety.

An awareness of the above scenario is essential in assessment. Although the situation is very different (Table 6.2), any perceived or measured reductions have to be related to a base which is meaningful in design model terms. What to do here will be covered later when dealing with particular action effects, but, in principle, there are two options:

(i) establish the assumed 28-day characteristic strength in the original design and make allowance for any subsequent changes (see Note 1 to Table 6.3);

(ii) take representative cores from damaged and undamaged regions in the structure (cores are usually taken in the direction of least lateral restraint) and integrate the resulting strengths into the relevant design models.

Table 6.3 is not the complete picture as far as mechanical properties are concerned in design. It is also necessary to consider strains and, in particular, the stress–strain curve. Figure 6.3 shows the normal schematic representation for concrete in compression, making recommendations for

Table 6.3 Summary of relationships between the mechanical properties for concrete, as assumed in current design [6.2] (Permission to reproduce extracts from the British Standards is granted by BSI)

f_{ck} (MPa)	12	16	20	25	30	35	40	45	50	55	60	70	80	90	Characteristic 28 d cylinder strength
$f_{ck.cube}$ (MPa)	15	20	25	30	37	45	50	55	60	67	75	85	95	105	Characteristic 28 d cube strength
f_{cm} (MPa)	20	24	28	33	38	43	48	53	58	63	68	78	88	98	Mean strength $= f_{ck} + 8$ (MPa)
f_{ctm} (MPa)	1.6	1.9	2.2	2.6	2.9	3.2	3.5	3.8	4.1	4.2	4.4	4.6	4.8	5.0	f_{ctm} = mean tensile strength $= 0.30 \times f_{ck}^{(\frac{2}{3})}$ for $f_{ck} \leq 50$
$f_{ctk,0.05}$ (MPa)	1.1	1.3	1.5	1.8	2.0	2.2	2.5	2.7	2.9	3.0	3.1	3.2	3.4	3.5	$f_{ctm,0.05} = 0.7 \times f_{ctm}$ 5% fractile
$f_{ctk,0.95}$ (MPa)	2.0	2.5	2.9	3.3	3.8	4.2	4.6	4.9	5.3	5.5	5.7	6.0	6.3	6.6	$f_{ctk,0.95} = 1.3 \times f_{ctm}$ 95% fractile
E_{cm} (GPa)	27	29	30	31	33	34	35	36	37	38	39	41	42	44	$E_{cm} = 22[f_{cm}/10]^{0.3}$, f_{cm} in MPa

Notes

1. Formulae are given for estimating mechanical properties at ages other than 28 d, with a dependence on cement type.
2. In the UK, for stress block calculations at the ULS, design compressive strength $(f_{cd}) = 0.85 f_{ck}/\gamma_c [\gamma_c = 1.5]$.
3. f_{ctm} is a direct concentric tensile strength. The link to splitting tensile strength is $f_{ct} = 0.9 f_{ct.sp}$. The link to flexural tensile strength is:
   $f_{ctm,fl} = \max\{[(1.6 - {}^h/_{100}) f_{ctm} \times f_{ctm}]$ where $h =$ the overall section depth in mm.
4. The different values of tensile strength are used in different verification procedures, e.g. cracking and bond, but the general design value (f_{ctd}) is given by 1.0 $f_{ctk,0.05}/\gamma_c$.
5. E_{cm} values are for quartzite aggregates. Values for limestone are 10% less, and 30% less for sandstone, but 20% higher for basalt.

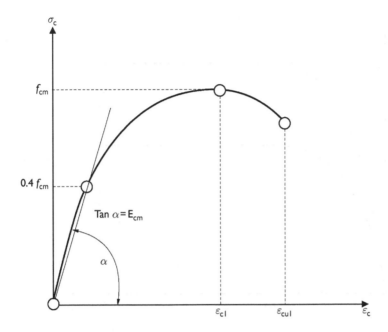

Figure 6.3 Representative stress–strain curve for concrete in compression used for structural analysis (Permission to reproduce extracts from the British Standards is granted by BSI)

the calculation of E_{cm} (Table 6.3 formula) for structural analysis; ε_{cu1} is taken as 0.0035 for concretes up to grade 50, and ε_{c1} is in the range of 0.0018–0.00245.

For the design of cross-sections, Figure 6.3 is idealised, the most common form being the one shown in Figure 6.4; f_{cd} is as defined in Note (2) to Table 6.3, with ε_{c2} and ε_{cu2} being 0.0020 and 0.0035 respectively.

In the context of assessment, stress–strain curves come into consideration for severe cases where expansive disruptive actions may not only affect the mechanical properties in Table 6.3 but also reduce the strain capacity in compression. It is also worth noting that BS EN 1992-1-1 contains a stress–strain curve for confined concrete, which could be useful in assessment, provided a realistic estimate could be made of the lateral compressive stress.

6.4.3 Mechanical properties for reinforcement and prestressing steel

The required characteristics for reinforcement are defined in CEN or National Standards, covering properties such as:

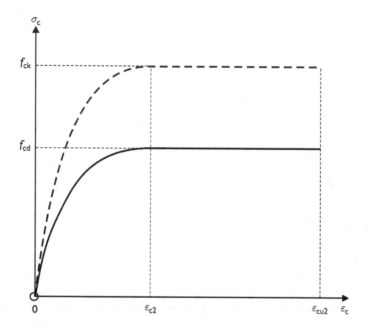

Figure 6.4 Parabolic–rectangular design stress–strain curve for concrete in compression (Permission to reproduce extracts from the British Standards is granted by BSI)

Yield strength	Bendability	Fatigue
Tensile strength	Bond	Tolerances
Ductility	Weldability	

Characteristic yield strength can be in the range 400–600 MPa, but usually at a characteristic value of $f_{yk} = 500$ MPa.

Typical stress–strain curves are shown in Figure 6.5 for hot rolled and cold worked steel. Different ductility classes are defined by setting limits for f_t/f_{yk} and ε_{uk} in Codes and Standards. For design, an idealised stress–strain curve is used. Commonly, this is bilinear, with a horizontal branch $f_{yd} = f_{yk}/\gamma_s$, where γ_s is the partial safety factor for reinforcement, normally 1.15. This horizontal branch starts at a strain of f_{yd}/E_s, where E_s is the elastic modulus.

A similar situation exists for prestressing steel, i.e. the required properties are defined in Standards for the different types of tendon. A typical stress–strain curve is shown in Figure 6.6, which shows that the 0.1 per cent proof stress $(f_{p0.1K})$ is used instead of the yield stress.

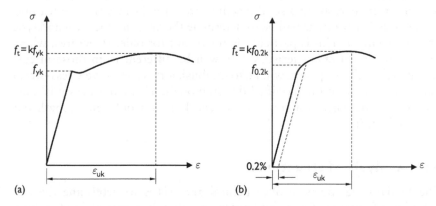

(a)   (b)

Figure 6.5 Typical stress–strain curves (a) hot rolled steel (b) cold worked steel (having a less well-defined yield stress, and therefore the 0.2 per cent proof stress is used) (Permission to reproduce extracts from the British Standards is granted by BSI)

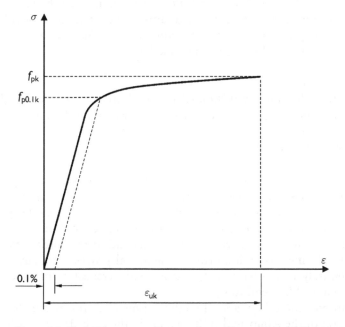

Figure 6.6 Typical stress–strain curve for prestressing steel (Permission to reproduce extracts from the British Standards is granted by BSI)

Most commonly in design, element strength is controlled by the reinforcement. In assessment also, this will be the case. Hence, the key features at the investigative stage will be to determine the amount, location and type of reinforcement. This is particularly important for older structures, where the reinforcement characteristics may well be different from those in current practice. The objective will be to establish a representative stress–strain curve, to be used in the modified design models. This latter action is most relevant to the corrosion situation, in checking not only on strength but also on ductility.

6.5 Analysis of structures

The fundamental purpose of structural analysis is to determine the distribution of internal forces and moments, due to imposed loads whose magnitude and distribution are laid down in Codes or Standards with the intention of generating maximum design values for these internal action effects. Analysis may also be used to evaluate stresses, strains and displacements.

A brief overall perspective of analytical methods is given in Section 5.3.3, with a bias towards assessment. At the detailed assessment stage (Figure 5.1), it is necessary to look at analysis in a little more detail, since the strategy in this book is to model and modify design approaches, and the modelling of the geometry and proportions of the structure becomes more acute.

In current design Codes [6.2], recommended approaches are:

(i) linear elastic behaviour;
(ii) linear elastic behaviour, with limited re-distribution based on specific assumptions, and some restrictions, e.g. not for column design;
(iii) plastic behaviour;
(iv) non-linear behaviour.

(i) and (ii) are by far the most common for most structures, particularly for frames and sub-frames in buildings, but with yield-line analysis (iii) making something of a come back for slabs. In undertaking this work, Codes give guidance on how to determine effective spans and effective flange width, and on how to deal with geometric imperfections and second-order effects where these are deemed to be important. The actual process of analysis is now almost entirely computer based, and a good perspective of modern methods is given by MacLeod [6.4].

Special mention should be made of analytical methods for bridge decks. The introduction of the Abnormal (HB) vehicle in the early 1950s, with its heavy indivisible point loads, promoted a move towards load distribution analysis based on orthotropic plate theory [6.5]. This was gradually replaced by grillage analysis in the 1960s. The current position is summarised in

Table 6.4 Current approaches to bridge deck analysis [6.6]

	Analytical approach	Example
Increasing complexity	Simple methods	Unit width strip method
	Linear numerical methods	Finite elements and grillage
↓	Upper bound plastic methods	Yield line analysis
	Non-linear numerical methods	Non-linear finite element analysis

Table 6.4. It is taken from reference [6.6], which looks at structural analysis in the context of assessment, and includes coverage of the guidance given in numerous Standards and Advice Notes issued by the Highways Agency. A more design-orientated treatment of analysis for bridges is provided by Clark [6.7].

The procedures outlined briefly above are part of an overall design package targeted at achieving acceptable levels of safety and serviceability (Section 6.3.3). In general, elastic-based design methods in Codes tend to be conservative, and the assumed nominal design loads may never be reached. The process contains many assumptions (Table 6.2), and real performance may be quite different. This can relate to mechanical and section properties, but a particular difficulty is with boundary conditions. Normally, supports are assumed to be either pinned or fully fixed, but reality lies somewhere in between, and is difficult to ascertain. This has to be borne firmly in mind in assessment work, by taking a close look at support conditions, including whether or not there is significant cracking at supports in continuous elements. The effectiveness of any bearing systems may also have to be considered; failures have occurred in the past where bearings have been too close to the edge of the support system in frames which were inherently flexible. Bearings designed to allow for rotation and/or longitudinal movement can malfunction, leading to unforeseen longitudinal forces, which can sharply reduce the bearing capacity of the support. Tolerances and construction method may also warrant consideration; e.g. for the Pipers Row car park described in Chapter 2, the distribution of loads around the perimeter of the columns was far from uniform as assumed in the design. A key input into structural analysis is that of flexural stiffness. In design, the elastic modulus [E] is taken as the mean value (Table 6.3 and Figure 6.3); the designer is looking for an average value, representative of the structure as a whole. The moment of inertia [I] is calculated on the assumptions of uncracked cross-sections and linear stress–strain relationships.

In assessment, this provides a starting point for analysis, and the key question is how much the effective EI value may have changed due to deterioration. Judgments on I values can only be based on the extent of cracking, particularly over supports, but also at midspan. Values for E can be more difficult to obtain. While tests may be done on cores taken from damaged and undamaged areas, an alternative approach is to relate E to estimates of effective concrete strength, as will be used in the following sections of this chapter, in calculating residual capacities at critical sections, while making some allowance for how much greater the mean strength might be, compared with the selected 'effective damage-related uniaxial strength'.

What might this all mean for structural analysis in detailed assessment? Much will depend on the nature and extent of the deterioration, since this will affect the depth and sophistication of the analysis. There are some basics however, as the analysis proceeds, and these would include the following:

Estimation of the actual imposed loads. In the author's opinion, this estimation should always be made. It relates to both items (2) and (3) in Table 6.2, and can have a direct influence on the γ_{F1} component of the partial safety factor for loads (Section 6.3.3).

Determination of the degree of fixity at the supports. This may be difficult to do in practice, but should be attempted nevertheless, since it can have an influence on the γ_{F2} component of the partial safety factor for loads. For determinate elements, generally there is no problem, although apparently simply-supported precast components often have some degree of fixity due to local details at the supports. Essentially, the objective is to establish whether the support system is closer to being pin-ended or fully fixed. Over the years, a great variety of details has been used both for bridges and buildings and hence the proximity to either end of the fixity scale can vary a great deal, even without any influence from deterioration. If there is excessive cracking due to deterioration at support sections in continuous elements, this will shift behaviour from fixity towards being simply-supported.

Selection of the value for moment of inertia [I]. The original design will almost certainly have been based on uncracked sections. It is suggested that this should also be the starting point for structural analysis in assessment. Only if the structural damage is severe (excessive cracking, spalling at critical sections, delamination, etc.) should reductions be considered as the analysis progresses.

Choice of values for modulus of elasticity [E]. Table 6.3 indicates the values likely to have been used in design, while showing that these are related to mean rather than characteristic strength (also see Figure 6.3). It is known that deterioration mechanisms such as ASR and frost action reduce the elastic modulus of the concrete and that this reduction can be disproportionately greater than that for compressive strength (see Tables 5.13 and 5.20). This may also be true for other aggressive actions such as sulfate attack, although less is known about the magnitude of these reductions (see Section 4.3.5, particularly Section 4.3.5.2 on thaumasite).

An obvious solution to this problem is to take cores from damaged and undamaged regions, and determine values for E from these. For severe and very severe conditions of frost attack (Section 5.6.3) this is virtually essential, since E values can be drastically reduced. This should provide an average value for E to be used in analysis.

Where testing is not feasible, but the original design and construction records are available, an approximate value for E can be obtained by working from Table 6.3. Knowing the original 28-day design strength, a value can be estimated for mean strength, while using Code equations for predicting strength gains beyond 28 days (Note 1 to Table 6.3). This allows values for E to be calculated for the undamaged concrete. For ASR, it is suggested that reductions in that value should be made in accordance with Table 5.13. For severe and very severe cases of frost damage, it is suggested that the reduction in E should be taken as 50–70 per cent, since an average value is required for the structure as a whole.

None of these is an exact science, and engineering judgement is required on all fronts in individual cases. It is important not to lose sight of the overall objectives in structural analysis in assessment which are:

1) To estimate how the distribution of internal forces and moments may have changed, due to deterioration;
2) To relate (1) to what was provided in the original design and to compare the relative values for safety levels (Section 6.3.3).

6.6 Overview of the effects of deterioration on the strength of elements and sections

In later sections of this chapter, proposals will be put forward on how to adapt design equations to allow for the effects of deterioration. These modifications will be based on a review of extensive experimental data, largely carried out during the CONTECVET programme [6.8–6.10], augmented by the work of Webster [6.11] and a report from the British Cement Association under the DETR PIT programme [6.12].

Most research on structural performance has been carried out on corrosion and ASR, although not always spread uniformly across all the relevant action effects, e.g. shear and punching shear are quite poorly represented as far as corrosion is concerned, and there are some practical difficulties in modifying the design models to calibrate with the experimental data. Nevertheless, a perspective emerges, and the object of this section is to put that across.

The approach is to take each action effect in turn, and to sub-divide in terms of individual deterioration mechanisms, wherever necessary.

6.6.1   Flexure

In general, experimental data indicate that flexural capacity is little affected by deterioration, unless the effects are severe, e.g. extensive cracking or spalling, or major reductions in concrete strength for elements that are over-reinforced. Most elements are under-reinforced, with capacity largely dictated by the reinforcement, provided that:

- the damage has not affected bond and anchorage;
- laps and splices remain operational; and
- minimum shear links are present.

For ASR, it is only when the free expansion is greater than about 6 mm/m, that reductions have been noted, generally no greater than 25 per cent. Clark [6.13] has estimated the flexural capacities of elements tested by various authors, by estimating restrained strains and converting these into steel stresses and forces. Calculation was on the basis of strain compatibility,

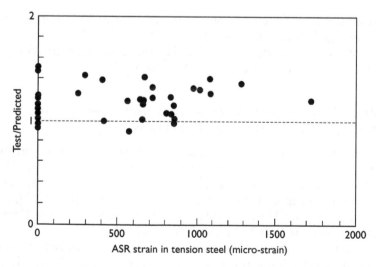

Figure 6.7 Effect of ASR on the flexural strength of reinforced concrete beams [6.13]

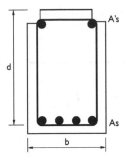

Figure 6.8 Modified section properties used for calibrating design model to allow for deterioration (mid-span bending) [6.12]

while allowing for ASR-induced pre-strains in the same way as for those due to conventional prestress. Figure 6.7 indicates the results obtained. It is stressed, however, that the method is only effective where the effects of the ASR expansion are limited as indicated above.

For corrosion damage, a similar picture emerges, again assuming that the damage is not severe. Even then, in using modified design equations, allowance can be made for the different possible effects of the deterioration, as indicated in Section 6.2, in relation to Figure 6.1. Verification of this was provided in reference [6.12], using experimental data obtained from reference [6.9].

The cross-section used in reference [6.9] is shown in Figure 6.8. In modifying the design equation, the following section modifications were made:

- the compression zone was partially reduced due to spalling at the corners;
- the areas of reinforcement used were based on measurements of average section loss;
- all other section and material properties remained unchanged (d, b, f_y f_{cu}).

The results of the analysis/calibration are shown in Figure 6.9.

The average test/predicted ratio for the ten control specimens used (no corrosion) was 1.053. If the assumptions made in allowing for deterioration were reasonable, and there were no other secondary effects, one might expect a similar test/predicted ratio, with some random scatter about this point. However, Figure 6.9 shows a downward trend as corrosion increases, although not directly proportional to the loss of rebar section. A possible reason for this lies in the measurement of the loss of cross-section, which is difficult to do and is undertaken after failure rather than before. It is also

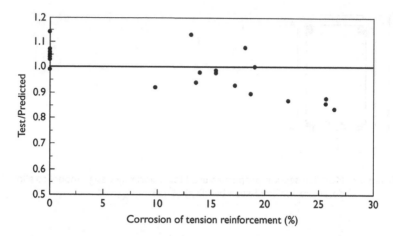

Figure 6.9 Results of analysis/calibration of design model allowing for deterioration [6.9, 6.12]

possible that loss of bond may have been a feature for the higher levels of loss of section. Having said that, the results demonstrate the viability of using modified design models for assessing flexural capacity – preferably with precise inputs on the physical deterioration.

The author has been unable to find comparable experimental data on beams affected by frost action, so the remarks which follow can only be speculative, and based on Section 5.6 in Chapter 5. Frost damage is most likely in planar elements such as slabs or massive structures such as dams, where the concrete is close to full saturation. Figure 6.10 is an example from a car park taken from reference [6.14].

If similar modelling to that described above for ASR and corrosion is attempted for frost damage, then a major factor will be the occurrence of surface scaling, where allowance for the loss of effective section would be made as was done in Figure 6.8 for corrosion. For internal mechanical damage, reductions in concrete strength could be made in accordance with Table 5.20, in modifying the basic model in Figure 6.1.

6.6.2 Compression

Compression comes into play mainly with columns or walls which can be subjected to pure axial load or a combination of axial load and bending. Experimental work on deteriorating columns is limited. Some work has been done with respect to ASR, e.g. Clark [6.13] and Chana and Korobokis

Figure 6.10 Typical example of freeze–thaw damage to an unprotected top deck of a car park, showing scaling of the surface [6.14]

[6.15], and tests on columns with corroded reinforcement were conducted as part of the CONTECVET programme [6.9].

There are two issues in detailed assessment terms, which, though related, can be considered separately. These are:

– section capacity;
– element behaviour, in terms of slenderness ratio and function in braced or unbraced frames.

6.6.2.1 Section capacity

In design, capacity is taken as the sum of that due to the concrete and the longitudinal reinforcement acting in compression, assuming that this is prevented from buckling by the presence of well-anchored links.

For ASR, it has been shown [6.15] that, for free ASR expansions of up to 4 mm/m, capacity is not reduced by more than that due to decreases in the compressive strength of the concrete. The question then is:

How should that decrease be estimated?

A popular approach is to work in terms of uniaxial compressive strength, ideally by taking cores from damaged and undamaged areas, and making comparisons. Should that be not possible, then Table 5.13 could be used, for conservative estimates. A further situation to consider is where the expansion has caused cracking, even spalling, in the concrete cover, with the possibility of only the core concrete remaining effective. Assuming that the restraint provided by the links remains effective, then the concrete will be subjected to triaxial stress, and advantage can be taken of Code equations to allow for this [6.2].

The situation for corrosion is broadly similar [6.9]. Further work by Clark [6.16] has shown that the longitudinal bars would buckle, rather than yield, only if the ratio of the link spacing to main bar diameter exceeds 44 and 32 for mild- and high-yield steel respectively. This assumes that the links remain effectively anchored – which may not always be the case for old structures – and are not themselves corroded.

The key features in estimating section capacity in compression are:

- obtaining a realistic and representative value for concrete compressive strength;
- checking on the nature and extent of the damage to the concrete cover;
- examining the links for signs of corrosion, while checking on the effectiveness of their anchorage.

6.6.2.2   Element behaviour

Columns may be slender or stocky. They may be contributing to the overall stability of the structure while acting as bracing members in frame action or as unbraced elements.

Codes [6.2] recognise all the possibilities, and make recommendations for slenderness ratios, effective lengths, etc. which reflect the different possible buckling modes. Simplifying assumptions, based on Eurocode 2, are given in reference [6.17], with follow-up recommendations for design. As part of this process, recommendations are given for load arrangements, while giving guidance on how to allow for geometric imperfections, and on how to cope with possible secondary effects in structures with flexible bracing systems.

While all of this is preliminary in preparation for calculating capacities of critical sections, there is also a primary role for compression members in contributing to overall stability. How, then, might this be affected by deterioration?

In the CONTECVET experimental programme on columns with varying levels of corrosion [6.9] it was noted that failure was initiated by cracking and spalling of the concrete cover, followed, in extreme cases, by failure of one or more corroded links. An increase on load eccentricity also occurred, due to the asymmetrical deterioration of the concrete – cracking and spalling occurred locally, and not uniformly on all four faces. This prompted the suggestion that an additional eccentricity be included in the analysis, in addition to that for geometric imperfections, and values for this were suggested [6.9].

With the above suggestion in mind, an assessment has then to be made of the real conditions for the compression element, in terms of end conditions, effective length and slenderness ratio, leading to a re-analysis for comparison with the original design. Central to all of this is an evaluation of the links, in terms of their anchorage and whether or not there is local corrosion. If the central concrete core remains intact, overall performance in terms of strength and stability is unlikely to be significantly affected, although the actual behaviour may be different from that assumed in design.

6.6.3 Shear

6.6.3.1 History and perspective

It is over 100 years since Morsch first proposed his truss analogy to explain shear behaviour. For design, he used a 45° truss with all the shear being carried by shear steel (links or bent up bars), and this was the basis for all Code design models until the 1960s. The year 1961 saw the publication of the so-called 'Stuttgart shear tests' [6.18] and the activities of the Shear Study Group in the UK, referred to by Regan [6.19]. Essentially, the research showed that the shear capacity of members with link reinforcement was greater than that could be explained by 45° truss action alone, and a concrete term was added to the shear link term to account for this. This practice of calculating the concrete and link contributions separately and then adding them continued in UK Codes for the next 30–40 years, e.g. reference [6.20].

Different UK Codes have different formulations for the concrete contribution, but are all essentially empirical, justified by fits with test data; there is no underlying rationale, as there is for flexure – making their adaptation for use where deterioration is involved rather difficult.

Slowly, a clearer picture emerged regarding member behaviour with no shear steel present; Taylor [6.21] pioneered this. He demonstrated that the contributions to shear resistance were as shown in Table 6.5.

Further work by Chana [6.22], involving the careful recording of strains in the three components listed in Table 6.5, broadly confirmed Taylor's findings (but with a slightly larger contribution from dowel action). Failure was initiated by dowel splitting cracks, suggesting that dowel action and aggregate interlock are inter-dependent – dowel action is essential for keeping the two components together and is therefore of vital importance for beams with no shear links. Moving this forward into a corrosion situation would suggest that cracking along the line of the tension reinforcement might pre-empt dowel-splitting at lower loads. All of this works also emphasises the importance of anchorage and bond of the tension reinforcement when considering the shear strength of deteriorating beams having no shear steel.

Table 6.5 Relative contributions to the concrete component of shear resistance [6.21]

Component	Contribution to concrete shear resistance (%)
Concrete compression zone	20 to 40
Dowel action of longitudinal tension reinforcement	15 to 25
Aggregate interlock along the main shear crack	30 to 50

Meanwhile, over the last few decades, design models for shear have continued to evolve. Instead of a fixed 45° truss, a variable truss method has come into being. This has been presented by Regan [6.23] and is now incorporated as an option in Eurocode 2 [6.2]. For bridges, this has recently been recommended for assessment purposes [6.24]. Canada and North America have led the evolution of the modified compression field theory, providing a rationale, representative of observed behaviour in shear [6.25, 6.26].

The reason for this preamble is to emphasise that the majority of Code design methods for shear are largely empirical, and, in assessing deteriorating elements, a viable alternative to simply modifying the traditional Code equations is to use alternative models. In this context, the principal ones are:

– the 45° truss approach, adding the concrete and link contributions;
– the variable truss approach, again adding the concrete and link contributions;
– the modified compression field approach.

Before considering which option to use in a particular case, it is necessary to look at what is known about the effects of deterioration on shear behaviour and strength, by reviewing available test data.

6.6.3.2 The effects of ASR on performance in shear

Structurally, there are two relevant components to the effects of ASR. First a loss in concrete strength. Second, a pseudo-pretress effect due to the expansive nature of the reaction, in the presence of both internal (reinforcement) and external (adjacent member) restraints to that expansion. The first is potentially detrimental to shear strength, in terms of directly affecting the concrete contribution, and indirectly, in extreme cases, in reducing bond and anchorage, and hence reducing the contribution from dowel action (Table 6.5). The second is potentially beneficial, provided that the restraints are generated, and the restrained expansion can be estimated.

Chana and Korobobis [6.27] conducted tests on beams with and without shear reinforcement, as did Cope and Slade [6.28]. In general terms, where links were present, ASR appeared to have little effect on shear capacity. For beams without links, reductions of 20–30 per cent in capacity were obtained where ribbed bars were used and of 15–25 per cent for smooth bars. However, the reduced capacities were still greater than Code predictions, using the fixed 45° truss approach. In his critical review of the structural implications of the effects of ASR, Clark [6.16] suggested that the shear capacity of reinforced concrete elements could be assessed by treating them as prestressed. In doing so, he recommended that the uniaxial concrete strength should be used, but that only 50 per cent of the ASR-induced prestress should be taken into account. Figure 6.11 shows the results that he obtained by this approach, giving a reasonable lower bound prediction to experimental data exhibiting some scatter, even though the predictions appear to be less safe, as the expansion increases. In summary, most work in the UK has involved making comparisons with predictions obtained from Code equations based on the 45° truss model, while adding the contributions from the concrete and the shear links. If an allowance is made for the reduction in uniaxial concrete strength due to ASR, in the concrete contribution to shear resistance, reasonable conservative estimates of residual shear capacity will be obtained, provided that the damage due to the action has not led to:

– loss of bond and anchorage in the tension reinforcement, and/or
– significant loss of concrete section due to cracking or spalling

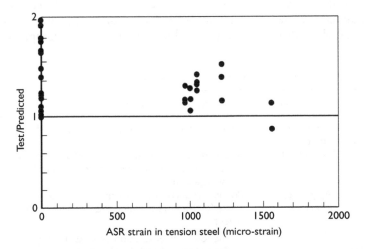

Figure 6.11 Effect of ASR on shear strength of reinforced concrete beams. Clark [6.16]

The author has been unable to find comparable test data on shear capacity where other types of expansive action are involved or for frost action. However, as for bending, common sense suggests that structural performance will be similarly affected as for ASR, and the approaches suggested here would be a useful starting point for assessment.

6.6.3.3   The effects of corrosion on performance in shear

With expansive actions such as ASR (6.6.3.2) the principal effects on shear capacity are a reduction in the uniaxial compressive strength of the concrete, and, in extreme cases, a reduction in the dowel action contribution to strength when the bond anchorage of the tension steel is affected. With corrosion, reductions in bond and anchorage can again be critical (as affected by the action of the expansive corrosion products). The other additional factor is loss of cross-sectional area, either of the main steel or of the shear links. Table 6.6 presents a brief summary of these effects.

Although corrosion is the major deterioration mechanism for structural concrete, there are less experimental data available in the public domain than for ASR. In this section, reliance is placed on the evaluations made in references [6.11] and [6.12], using the information contained in references [6.9] and [6.29]. More recently, reference [6.24] lists a series of reports on work at the University of Westminster (not generally in the public domain), and reference [6.30] focuses on the key feature of what constitutes effective anchorage, using finite-element methods to make comparisons with experimental data.

The experimental work described in the CONTECVET Manual [6.9] was carried out on beams containing deformed bars and links. The shear failures obtained are shown in Figure 6.12. Figure 6.12(a) shows a dominant shear crack with further cracking along the line of the compression

Table 6.6 Summary of the possible effects of corrosion on shear capacity

	Component	Possible effects of corrosion
Concrete contribution	Concrete compression	Dimensions of concrete section reduced by spalling.
	Dowel action	Reduced due to loss of steel cross-section and to the presence of longitudinal cracks reducing anchorage bond.
	Aggregate interlock	Possible reduction due to the tensile steel being less effective in restraining the opening of shear cracks.
Link contribution		Reduction in cross-section of links, usually preferentially at bends, without necessarily reducing their contribution to shear proportionally.

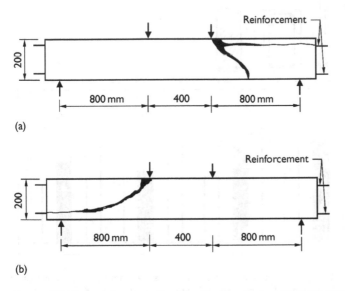

Figure 6.12 Shear failures obtained by Geocisa, in the CONTECVET project [6.9]

reinforcement. Figure 6.12(b) again shows a dominant shear crack but with a clear influence due to bond failure.

In this project, the loss of rebar section due to corrosion was measured. In making comparisons with Code equations then, various assumptions were made regarding the loss of concrete section due to cracking and spalling. The predictions obtained are shown in Figure 6.13, when adapting the standard design method in Eurocode 2 [6.2, 6.17]. It may be seen that a consistently conservative estimate was obtained only when calculations were based on the assumption that all the cover to the links had spalled.

The approach illustrated in Figures 6.12 and 6.13 is useful in gaining a perspective of the possible range of shear capacities, using one particular set of design equations and assuming that the loss of rebar cross-section is known. However, the scope is limited to the use of deformed bars in the presence of links. To expand the coverage, it is necessary to consider:

– cases were no shear steel is present;
– cases where plain bars are used as main reinforcement; and
– the possible influences of

(i) cover/diameter ratios,
(ii) loss of anchorage bond to some extent, and
(iii) any interaction between main reinforcement and links in providing alternative modes of shear resistance.

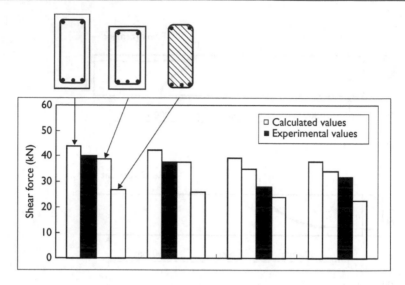

Figure 6.13 Comparisons between experimental shear capacity and predictions made by the standard method in Eurocode 2, based on different assumptions on loss of concrete section CONTECVET [6.9]

The analytical work by Webster [6.11], based on the limited experimental results in references [6.9] and [6.29], provides some guidance on the above issues. In evaluating the data, Webster used the models contained in BS 8110 [6.20], EC2 [6.2] and the Canadian Code [6.31] (which is based on the modified compression field theory), making comparisons between them and putting forward suggestions on which was best for different assessment situations. The main conclusions from his work are as follows:

1) *Element having no shear reinforcement*

   (a) The cover to bar diameter ratio (c/d) appears to have a significant beneficial effect on shear capacity, once corrosion has occurred.
   (b) Dowel action is an important factor, as its influence in a corrosion situation will depend on the anchorage bond conditions which then obtain.
   (c) In evaluating the concrete contribution to shear resistance, it is necessary to determine the effective area of tension reinforcement.
   (d) Allowing for the effects of corrosion on the concrete contribution requires the normal Code equations to be modified to allow for the c/d ratio and the ratio of corroded bond strength to that assumed in design.

2) *Elements having shear links*

(a) By themselves, links contribute to shear resistance, making a contribution which can be added to that from the concrete. Conservatively, this may be done by reducing their area, based on an assumed level of corrosion. In reality, corrosion of links is likely to occur preferentially at bends, and the remaining legs will still function to some extent.

(b) Even where significant loss of concrete cover has occurred, the presence of links enhances residual bond strength and is therefore likely to contribute to shear strength.

3) *Elements having plain bars as main tension reinforcement*

Beams reinforced with plain bars can achieve higher shear capacities than similar beams with ribbed bars. This phenomenon also seems to occur in the presence of corrosion and is attributed to the beam acting as a shallow tied arch, provided that the end anchorages remain intact. Webster [6.11] developed formulae to account for this, which gave reasonable predications when compared with the experimental data from Daly's work [6.29].

4) *The relevance and applicability of the different Code models*

Webster [6.11] compared the predictions from the BS 8110, EC2, and CSA A23.3 models with the Daly's experiments [6.29] for beams having no links, and with both Daly and CONTECVET [6.9] for beams with links.

Table 6.7 shows the results of these comparisons for beams having no shear links; a similar table was produced for beams with shear links.

In analysing the results, two factors are of interest:

- the mean test/predicted ratio, as a measure of how good the predictions are;
- the relative values of the test/predicted ratios for the control beams (no corrosion) compared with those for the corroded beams having different percentages of corrosion. This gives an indication of whether or not all the significant differences between uncorroded and corroded beams have been accounted for. In doing so, an ideal target for the models is to have the same mean test/predicted values for both, thus reproducing the safety margins implicit in the original design.

Table 6.7 Comparison of BS 8110, EC2, and CSA A23.3 models for calculating the shear strength of corroded beams having no shear links. Webster [6.11]

Beam	Corrosion (%)	Vtest (kN)	BS 8110 (mod)		EC2		CSA A23.3	
			Vpred (kN)	Vtest/ Vpred	Vpred (kN)	Vtest/ Vpred	Vpred (kN)	Vtest/ Vpred
TS12/0	0.0	56.9	44.3	1.28	52.8	1.08	43.2	1.32
TS12/1	4.3	48.6	39.7	1.23	51.9	0.94	44.5	1.09
TS12/2	21.5	39.0	33.5	1.16	48.5	0.80	44.2	0.88
TS12/3	21.5	45.8	33.5	1.37	48.5	0.94	42.2	1.09
TS24/0	0.0	53.5	43.2	1.24	51.3	1.04	41.1	1.30
TS24/1	4.5	51.6	43.2	1.19	50.4	1.02	40.9	1.26
TS24/2	14.4	50.0	40.9	1.22	48.5	1.03	39.5	1.27
TS24/3	21.5	46.2	36.9	1.25	47.1	0.98	39.2	1.18
TS36/0	0.0	47.9	42.1	1.14	49.6	0.97	39.5	1.21
TS36/1	3.8	51.2	42.1	1.22	49.1	1.04	38.1	1.34
TS36/2	14.1	49.7	42.1	1.18	47.0	1.06	36.6	1.36
TS36/3	17.9	48.7	41.3	1.18	46.3	1.05	36.2	1.35
Control beams			Mean	1.22	Mean	1.03	Mean	1.28
			St. dev.	0.123	St. dev.	0.075	St. dev.	0.057
			C.o.V	6.16%	C.o.V	5.56%	C.o.V	4.47%
Corroded beams			Mean	1.22	Mean	0.99	Mean	1.20
			St. dev.	0.061	St. dev.	0.082	St. dev.	0.158
			C.o.V	4.98%	C.o.V	8.32%	C.o.V	13.19%
All beams			Mean	1.22	Mean	1.00	Mean	1.22
			St. dev.	0.075	St. dev.	0.061	St. dev.	0.141
			C.o.V	4.99%	C.o.V	7.68%	C.o.V	11.59%

In making these comparisons, Webster used the EC2 and A23.3 models without modification, except for inserting reduced reinforcement areas to represent the level of corrosion in different test specimens. On the other hand, for the BS 8110 model ($45°$ truss), he introduced a modifying factor to represent possible reductions in the area of effectively anchored tension steel.

In Table 3.8 in BS 8110 [6.20], the design concrete shear stresses are based on an equation which contains the term

$$\left(\frac{100A_s}{bd}\right)^{1/3}$$

where A_s = area of effectively anchored longitudinal tensile steel
b = width of the section
d = effective depth to the tension reinforcement

Webster's proposal was to replace A_s by $A_s$eff, where $A_s$eff is the effective area of the tension reinforcement in the corroded beams. In doing so, he considered three variables: tension reinforcement corrosion, c/d ratio and corroded bond strength. Based on his general observations of the experimental data, his proposal was to use the expression

$$\frac{A_s\text{eff}}{A_s} = 0.8 \left(\frac{c}{d}\right)^{1/3} \frac{f_{b.corr}}{f_b}$$

where

A_s.eff, A_s, c, and d are as defined above

f_b is the bond strength of the uncorroded tension reinforcement, calculated in accordance with the Code

f_b.corr is the corroded bond strength of the tension reinforcement, calculated as indicated in Section 6.6.5.

This modification was used in producing Table 6.7 It may be seen that the mean values using this approach are identical for both the control and corroded beams, with a smaller coefficient of variation for the latter.

The predictions using the EC2 model give the best mean ratios for both the control and corroded beams, but with a higher coefficient of variation for the latter. A few of the predictions are marginally on the unsafe side. The CSA A23.3 approach produces safer results, but with a significant increase in the coefficient of variation for the corroded beams.

In general, Table 6.7 indicates that all three models can be used for predicting shear strengths of beams having no shear links, while suggesting that all could be marginally improved if more test data were available.

The corresponding comparisons for beams having shear links paint a broadly similar picture, but with mean test/predicted ratios closer to 1.0 and higher coefficients of variation. In fact, for the EC2 model, the mean test/predicted ratio for the corroded beams was less than 1.0, suggesting that some modifications may be required. There was greater variability in the predictions generally where shear links were present; possibly this is due to variations in the amount of corrosion in the links and depends on where that corrosion occurs.

There remains the question of beams reinforced with plain bars. As suggested earlier in conclusion four to Webster's work, such beams can exhibit higher shear capacities than similar beams with ribbed bars, and this may be due to tied-arch action, as illustrated in Figure 6.14.

For this action to take place the anchorage of the tension reinforcement must remain sound, ideally sufficient to yield the steel. Conventionally, shear capacity relies on bond strength to transfer forces in the tension steel to the concrete. As bond strength reduces, e.g. with plain bars, more so with corroded plain bars, the potential for force transfer also reduces until plain sections no longer remain plain. The alternative load-carrying

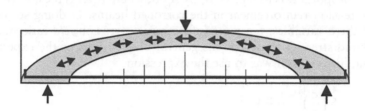

Figure 6.14 Schematic illustration of tied-arch action for beams with plain bars

mechanism shown in Figure 6.14 then comes more into play. In reality, a mix of conventional load transfer and tied-arch action will obtain, with a move towards the latter as the possible effects of corrosion become more pronounced.

This phenomenon has also been observed by Cairns, [6.32] in testing beams where the tension reinforcement was exposed over varying lengths, i.e. no effective bond. For the longer exposed lengths higher shear capacities were obtained, and Cairns suggested arching action to explain this. Webster developed this approach in his thesis and made comparisons with the experimental data from Daly's work. In doing so, he demonstrated that the arching action component of the shear resistance could vary between 40 per cent when full bond obtained to 100 per cent with no bond, but with sufficient anchorage in place to yield the tension steel.

Arching action is introduced here to suggest a possible way forward for beams with plain bars, where the normal force-transfer models in Codes produce capacities less than those theoretically required. Conventional Code models will generally produce conservative answers, and, if full anchorage can be guaranteed, better estimates can be obtained.

6.6.4   Punching shear

Here, we are dealing with planar rather than linear elements. The brief description of the collapse of the Pipers Row car park, given in Section 2.5.1, indicated that a combination of factors led to the failure, with the dominant deterioration mechanism being frost action. The extent and depth of the resulting deterioration led to a loss of bond and anchorage in the tension reinforcement, leading to a critical reduction and imbalance in the internal forces, which normally provide resistance to punching shear.

There is a lack of test data which would permit an evaluation to be made of the validity of current design models in assessing residual capacity where deterioration has occurred. Some speculation is feasible by looking at the nature of the design models and attempting a translation from what has been learned from the conventional shear situation.

Considering Eurocode 2 [6.2, 6.17], weight is rightly put on the establishment of a critical shear perimeter and on how to deal with load eccentricity. Both these features would also figure prominently in assessment in checking the as-built conditions; this would provide a realistic estimate of the distribution of the applied shears.

Probably the most common situation is slabs having no shear steel. In this case, the Code gives an expression for calculating a design punching shear resistance below which no shear steel is required. A follow-up expression then permits additional resistance to be provided by means of shear steel. Included in the expression for the basic resistance are the following:

- f_{ck} the characteristic concrete cylinder strength. In fact, $(f_{ck})^{1/3}$ is used.
- a term which takes account of the amount of orthogonal tension reinforcement within a specified distance from the column faces.

This gives pointers on what to look for in an assessment situation. If the nature and extent of the deterioration is such that internal mechanical damage has occurred (e.g. frost, ASR, sulfate attack) then reduced values for f_{ck} will have to be used, as indicated in Section 6.3.2 (although, for ASR, limited test data [6.33] would indicate that the prestressing effect of the expansion counters any reduction in concrete strength, provided that the free expansion is less than about 6 mm/m). If the internal mechanical damage associated with surface scaling and/or cracking is severe, then bond and anchorage of the tension reinforcement could be affected. Bond and anchorage should be checked when corrosion has occurred, since the function of this steel is fundamental to the equilibrium of the internal resistance mechanism to punching.

6.6.5   Bond and anchorage

6.6.5.1   Background

Design equations for bending, shear and axial load are based on the assumption that strains in the reinforcement and adjacent concrete are the same, i.e. that perfect bond exists. Codes require bond strength to be checked, and, while formulae for bond stress are given, common practice is to specify ultimate anchorage bond lengths and lap lengths in tension and compression as multiples of bar size. Table 3.27 in BS 8110 [6.20] is a typical example of this. The practice of checking local bond strength between flexural cracks has now largely been discarded.

In general, reinforcement can consist of plain or deformed bars. Current practice is based largely on the use of ribbed deformed bars, of which there can be different types. In the past, plain bars were more common, and square twisted deformed bars were also used. Plain bars rely on adhesion and friction for bond, and deformed bars rely additionally on mechanical interlock. Values for bond in the literature depend on the type of test used (generally either pull-out tests or using specimens designed to simulate the stress conditions in real structural elements), and there is usually considerable scatter in any given population. Plain bars fail by being pulled out through the concrete causing little local damage, whereas ribbed bars tend to fail as a result of longitudinal splitting cracks. As might be expected, deformed bars generate higher bond strengths.

There is also sufficient data available to indicate what parameters affect bond strength in practice. A summary of these due to Webster [6.11] is given in Table 6.8. The influence of all these individual factors is not treated explicitly in Codes, where design for ultimate bond strengths are in the format:

$$f_b = \beta\sqrt{f\text{cu}}$$

where

   f_b is the ultimate bond strength
   f_{cu} is the concrete compressive strength, generally taken as the characteristic value (Table 6.3)
   β is a coefficient dependent on the bar type.

Table 6.9 shows the β values to be used in BS 8110 [6.20] for different types of bar, including fabric. These values contain a partial safety factor of 1.4, and it is assumed that minimum links are provided. While there is a great deal of experimental work on bond in the literature, the basis for the bond clauses in BS 8110 comes primarily from the work of Reynolds [6.34] and Tepfers [6.35].

It is the author's opinion that bond and anchorage are of great importance in assessment work where deterioration has occurred, and requires more detailed attention than that given to it in conventional design. There are two principal reasons for this:

(1) the need to assess what the deterioration in general, and corrosion in particular, actually does to bond behaviour, and the impact that it might have on how the design equations are used for assessment;
(2) the need to attempt an evaluation of the parameters in Table 6.8, most of which provide beneficial effects on bond strength above those implicit in conventional design equations.

Section 6.6.5.2 presents a review of what is known about this, with emphasis on corrosion. This is followed by brief coverage of other

Table 6.8 The influence of different parameters on uncorroded reinforcement – Webster [6.11]

Parameter	Effect on bond strength	Applicable to	
		Plain	Ribbed
Cover to bar diameter (c/d) ratio	The bond strength increases with increasing c/d ratio up to a limiting value of approximately 2.5, above which no enhancement occurs. At higher covers there is a tendency for bars to pull out rather than split the concrete.		✓
Links	Links provide confinement to the concrete surrounding the longitudinal bar and intersect potential crack planes. As such, higher bond strengths can be achieved.		✓
Bar location	Bars near the top of members tend to have lower bond strengths than bottom cast bars. The concrete surrounding top-cast bars is not as well compacted as that at the bottom of a member. In addition, plastic settlement can lead to reductions in bond strength.	✓	✓
Concrete tensile strength	The bond strength increases with increasing tensile strength of the cover concrete for low c/d ratios.	✓	✓
Bar embedment length	The further that the bar is embedded in concrete beyond the point at which it is required, the higher the bond forces which can be resisted (up to yield or slip).	✓	✓
Applied normal stress	Applied stress such as that at member supports provides a confining action and limits splitting cracking. This increases bond strength.	✓	✓
Dowel action	Dowel action, due to transmission of shear forces, is likely to increase the possibility of longitudinal splitting and thus reduce bond strength.	✓	✓

Table 6.9 Recommended values for the bond coefficient β (BS 8110) [6.20]

Bar Type	β	
	Bars in tension	Bars in compression
Plain bars	0.28	0.35
Type 1 deformed bars	0.40	0.50
Type 2 deformed bars	0.50	0.63
Fabric	0.65	0.81

deterioration mechanisms where the principal effects are on the properties of the concrete.

6.6.5.2 The effects of corrosion on bond and anchorage

While there is quite a lot of test data available in the literature, it does suffer from a lack of consistency and realism, in terms of the following:

- different bond tests have been used, ranging from pull-out to beam tests;
- different corrosion rates have been used. Accelerated corrosion rates are common, producing a more liquid type of corrosion product, not representative of practice;
- many ratios of cover to bar diameter (c/d) are significantly higher than those used in practice;
- some of the test data relates to specimens where no cracking has occurred due to corrosion. At this stage, the corrosion products appear to add extra confinement to the bars, and higher bond strengths have been recorded than those for uncorroded bars. It is only when cracking occurs the bond strength tends to reduce.

The nett effect of all these is even more scatter in the test data compared with the uncorroded situation. Nevertheless, some general trends can be detected and Tables 6.10 and 6.11 are a compilation of these, again due to Webster [6.11]. The important point from Table 6.10 is the need to distinguish between the pre-cracking and post-cracking phases when corrosion has occurred. Table 6.11 examines the effect of the different parameters that can affect bond strength in practice and, when compared with Table 6.8, brings out differences between the corroded and uncorroded situations.

Plainly, some of these parameters are more important than others, in defining differences between the uncorroded and corroded cases. To use the information in Tables 6.8–6.11, it is necessary to isolate the key parameters and try to define their influence when modifying design approaches for

Table 6.10 Different phases due to corrosion, in terms of their influence on bond strength

Phase	Typical behaviour
Uncorroded	Behaviour is as assumed in design codes.
Pre-cracking	Expansive corrosion products are resisted by the surrounding concrete. The corrosion induces extra confinement to the bar. Light rusting on the bar surface increases the frictional resistance. The two together combine to increase the bond strength. This increase can typically be up to 1.5 times the uncorroded value.
Cracking	When the first crack appears, much of the confinement is lost and there is a drop in bond strength from the pre-cracking peak. Plain bars appear to exhibit a larger drop than ribbed bars. Bond strengths in the region of 0.9 to 1.2 times the uncorroded strengths have been observed in tests.
Post-cracking	The bond strength has been observed to reduce with increasing corrosion. As the ribs of deformed bars deteriorate, there is little difference between them and plain bars. Some tests have shown the residual bond strength to be 0.15 times the uncorroded values at 8% corrosion. However, other tests have shown the residual to be 0.6 at 25% corrosion (but at a lower corrosion rate).

Table 6.11 Effects of various parameters on bond strength – corroded reinforcement (compare with Table 6.8 for uncorroded reinforcement)

Parameter	Effect on bond strength
Cover to bar diameter (c/d) ratio	An increase in c/d generally increases the time to cracking, and thus time to loss of bond strength. An increase in the c/d ratio appears to increase bond strength at lower levels of corrosion, but bond strengths begin to converge at higher levels of corrosion.
Links	Links appear to offer greater benefit to corroded members than uncorroded ones. In tests 70 to 80% of the uncorroded bond strength was maintained when corroded with links present, compared to 20 to 30% when they were not.
Bar position	Once cracking has taken place, both top and bottom cast bars have similar bond strengths. That is, bottom cast bars suffer a greater proportional reduction in bond strength.
Concrete tensile strength	The bond strength appears to increase with increasing tensile strength.
Corrosion rate	Increasing the corrosion rate initially leads to increases in bond strength. Further increases in corrosion rate lead to reductions in bond strength for the same amount of corrosion.
Bar type	At first cracking, ribbed bars show a small drop in bond strength than plain bars. Post-cracking, ribbed bars showed a larger drop. Ribbed bars appear to require less corrosion to cause cracking than plain bars.

assessment. Dealing first with the situation where no links are present, the relevant parameters are:

– cover to bar diameter ratio (c/d);
– concrete tensile strength;
– amount of corrosion and corrosion rate; and
– whether or not cracking due to corrosion has occurred.

The type of bar would appear to be less influential when corrosion has occurred, but may also warrant consideration. The contribution of links can then be considered separately.

In sifting through the test data available to him, Webster [6.11] focused mainly on the results from the CONTECVET programme and from the University of Birmingham [6.36] as being the most consistent and realistic in deriving expressions for assessment, which took account of the main parameters listed above. Both investigations were conducted at corrosion rates reasonably representative of what might occur in practice, and both produced broadly similar results. This is illustrated in Figure 6.15, where R_b is the ratio of corroded to uncorroded bond strength, and corrosion percentage represents the loss of area in the longitudinal steel. The two corrosion rates are as shown in the legend, and give broadly similar results, while being mindful of scatter in any test data on bond.

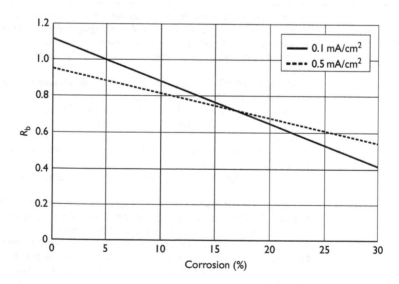

Figure 6.15 The effect of the loss of main reinforcement section on the ratio (R_b) of corroded to uncorroded bond strength ([6.9] and [6.36])

In analysing the test data, Webster [6.11] used the work by Tepfers [6.5] as a base (which includes terms for (^{C}/d) and concrete tensile strength) and based on linear regression analysis, derived a modification factor for the influence of corrosion, which also accounted for the amount of corrosion and corrosion rate. The resulting lower bound expression, corresponding to the 5 per cent characteristic, is as follows:

$$f_b = (0.31 - 0.015A_{corr})(0.5 + c/d)f_{ct} \tag{A}$$

where,

f_b is the corroded bond strength (N/mm^2)
A_{corr} is the loss in area of the corroded bars (mm^2)
(c/d)   is the cover to bar diameter ratio
f_{ct} is the concrete tensile strength, splitting (N/mm^2)

Figure 6.16 indicates that this gave a reasonable lower bound to the test data, for ribbed bars with no links. There is always a danger in practice in using this type of formulation based on relatively limited data, but the expression seems to account for the key variables quite well.

For corroded ribbed bars, there remains the question of what additional contribution to bond strength can be provided by the presence of links. There is clear evidence from test data, without fully understanding the mechanism involved, that the presence of links can make a significant

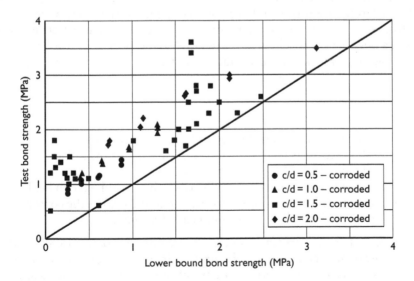

Figure 6.16 Comparison between predicted lower bound bond strength and test bond strength for corroded ribbed bars with no links

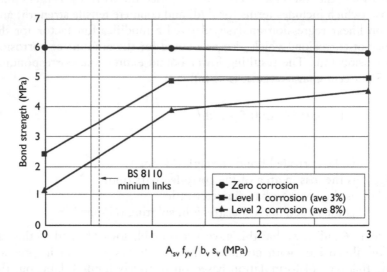

Figure 6.17 The influence of links on bond strength of corroded ribbed bars [6.9]

contribution. Figure 6.17 is a plot from the CONTECVET data for two different levels of corrosion, where the horizontal axis gives a measure of the link content. However, the data is limited, and the increase is not linear.

BS 8110 [6.20] allows an increase in the bond strength of uncorroded bars where minimum links are present of $0.2 \sqrt{f_{cu}}$. From the limited CONTECVET test data this would appear reasonable for the levels of corrosion tested, with a lower bound value of 1 N/mm². For higher percentages of links, a possible approach might be to develop the work of Reynolds [6.34], who suggested an enhancement due to links of

$$ f_{b.link} = \frac{k.As_v}{S_v.d} $$

$f_{b.link}$ is the additional contribution to bond from the links
A_{sv} and S_v are, respectively the area and spacing of the links
d is the diameter of the main tension steel
k is a coefficient. For the corrosion case, this would have to be derived from text data, which is currently insufficient, and therefore the conservative Code approach given above, is recommended at this time.

For plain bars, experimental data is limited to one source, [6.36] so that only indicative conclusions can be drawn. It would appear that the (c/d) ratio is significant when corrosion has occurred, unlike the corresponding case for uncorroded bars; conceivably, this is due to the irregular corrosion products producing some form of interlocking mechanism analogous to that for ribbed bars. Using the same approach as that for deriving expression (A) for ribbed bars, i.e. the Tepfer's approach [6.35], and regression analysis of the test data to give a 95 per cent characteristic value, leads to expression (B).

$$F_b = 0.36 - 0.11A_{corr}(0.5 + (c/d))f_{ct} \qquad (B)$$

where the terms are as defined for expression (A). The fit that this produces is shown in Figure 6.18.

In fact, this predicts strengths higher than the corresponding lower bound expression for ribbed bars (expression (A) and Figure 6.16), possibly due to the lower scatter in the test data. Bond and anchorage are fundamental to structural performance when cracking due to corrosion has occurred. This explains why this Section 6.6.5.2 has devoted so much space to the analysis of available test data, while attempting to use that to modify Code-based design equations for use in assessment. The data is limited, but the trends are right. Using this approach in practice requires a measure of engineering judgement and the gathering of as much detailed information as possible, e.g. on the extent and location of the cracking and whether or not the as-built anchorage and lap lengths do in fact comply with Code detailing rules. The presence of links is also important in enhancing bond strength, as is the beneficial effects of transverse confinement due to support reactions.

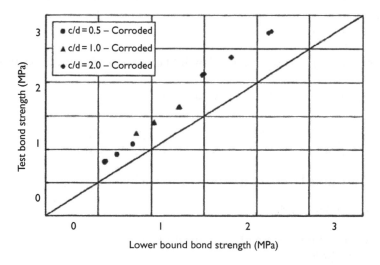

Figure 6.18 Comparison between predicted lower bound bond strength and test bond strength for corroded plain bars with no links [6.36]

6.6.5.3 The effects of expansive mechanisms on bond and anchorage

The author's experience relates mainly to ASR and frost action, using the CONTECVET manuals as a basis [6.8, 6.10].

Much will depend on the extent and magnitude of the damage, in terms of cracking or spalling. In severe cases where significant loss of cover has occurred, bond and anchorage will be reduced (as for corrosion) and performance will depend largely on the type of bar, the ratio of cover to bar diameter, and whether or not links are present.

For less severe cases, the major contribution to strength reduction will come from reductions in the mechanical properties of the concrete, as used in Code-based equations for bond. For ASR, a uniaxial compressive strength should be obtained from cores. Table 5.20 gives corresponding values for frost action, including values for bond strength derived from laboratory tests.

In practice, it is sensible to distinguish between cases where transverse links are present and where they are not. As an example, for ASR, the following is suggested.

(a) where minimum links are present, and cracking is not severe.

Reduce bond strength, as calculated by conventional Code equations as follows:

Restrained expansion (mm/m)	0.5	1.0	2.5	5.0	10.0
Reduction factor on uniaxial compressive strength	0.95	0.80	0.60	0.60	

(b) where no links are present, which would cross potential crack locations

(i) Reduce concrete tensile strength as follows:

Restrained expansion (mm/m)	0.5	1.0	2.5	5.0	10.0
Reduction factor		0.85	0.75	0.55	0.40

This gives a reduced tensile strength, as affected by ASR ($f_{ct.asr}$)

(ii) For ribbed bars

Ultimate bond strength f_b is given by
$f_b = \alpha(0.5 + c/d) f_{ct.asr}$
where c/d is the ratio of cover to bar diameter; $f_{ct.asr}$ is obtained as in (b) (i) above

Table 6.12 Values for α and β in calculating bond strength in ASR-affected elements with no links.

Bar location	α	β
Corner and top	0.30	0.33
Corner	0.43	0.47
Top	0.43	0.47
Elsewhere	0.60	0.65

α is a coefficient, given in Table 6.12.

(iii) For plain bars

$$f_b = \beta f_{ct.asr}$$
Where β is a coefficient given in Table 6.12.

Table 6.12 recognises that bond strength can very depending on bar location. This is also recognised in Codes where different design values are given for bond conditions defined as 'good' and 'other.' 'Good' corresponds to 'Elsewhere' in Table 6.12.

6.7 Cracking

Cracking is a variable phenomenon, and the possible causes are legion (Section 4.2.2). In assessment work, the norm would be to plot crack patterns (e.g. Figure 5.2) and to classify them (e.g. Tables 5.1 and 5.2), while identifying the dominant causes, since this will determine the nature and timing of any remedial action, possibly driven by appearance, serviceability and functional concerns. The emphasis in assessment is on physical observation and measurement.

However, cracks can act as locators, triggers and accelerators for any effects of deterioration and can also produce synergetic effects as indicated in Chapters 4 and 5. Where relevant, the structural effects or cracks were covered earlier in this chapter, e.g. on bond, anchorage and shear.

One possible relevance of cracking in detailed assessment is to cases where corrosion is the main concern, and cracking has not yet occurred, but chloride or carbonation profiles indicate a significant risk. The owner or assessor may wish to know if and when cracking will occur. This requires knowledge not only of chloride or carbonation profiles and the critical conditions for corrosion to start, but also of potential corrosion rate, as affected by local and internal micro-climates. This is not easy, since all these factors can be highly variable in themselves and can differ over different timescales.

Further, when corrosion starts, and corrosion products begin to form, there is the further complication due to some of the products diffusing into the concrete before expanding to cause radial pressure. Key questions then are:

(a) How much of the products disperse into the concrete, a function of the pore structure, without inducing stresses in the concrete?
(b) What radial pressure is necessary to cause cracking and how can we estimate it?

Attempts have been made to answer both these questions mainly via laboratory work and the development of analytical models. These result in equations to predict how much of the corrosion products disperse into the concrete and to calculate the radial pressure needed to cause cracking. From this, by introducing a value for corrosion rate, an equation is obtained for estimating the time to cracking.

Basic variables in doing this include:

- thickness of the concrete cover
- bar size
- tensile strength of the concrete
- rust occupying a greater volume than the original reinforcement
- the pore structure of the concrete, which is dependent on:

    water/cement ratio; degree of compaction; degree of hydration of the cement; fluidity of the rust products; presence of any fine cracks in the vicinity of the reinforcing bars – none of them easy to quantify in an assessment situation.

In reviewing available test data, Webster [6.11] concluded that the radial expansion of the rust into the concrete pore structure was significantly greater than the radial expansion required to induce cracking stresses. This led to his proposal that only a simplified method could be justified in estimating the time to cracking, since the major factor (rust dispersal) was difficult to quantify. Both phases of rust development appeared to be dependent on the ratio of concrete cover to bar diameter (c/d), and Webster developed a simple expression for bar section loss needed to cause corrosion-induced cracking of the form.

$$\delta_{cr} = k_1.c/_d \ (\%)$$

Analysis of the experimental data suggested that k_1, was of the order of 0.5. This then leads to an expression for estimating the time to cracking in years of the form

$$t_{cr} = \frac{k_1.c/d}{11.61I_{corr}}$$

Where I_{corr} = corrosion rate (μ A/cm^2)

The above proposal may appear to be hugely simplistic, but it is difficult to justify anything more elaborate unless a major effort is made to investigate the nature of both the corrosion products and the concrete pore structure.

References

6.1 British Standards Institution (BSI). *Eurocode – basis of structural design.* BS EN 1990: 2002. BSI, London, UK. 2002.

6.2 British Standards Institution (BSI). *Eurocode 2: Design of concrete structures – Part 1–1: General rules and rules for buildings.* BS EN 1992-1-1: 2004. BSI, London. 2004.

6.3 International Standards Organisation (ISO). *General principles on reliability for structures ISO 2394.* Available from BSI, London, UK.

6.4 MacLeod Iain A. *Modern structural analysis: modelling process and guidance.* Thomas Telford Ltd., London. 2005.

6.5 Rowe R.E. *Concrete bridge design.* CR Books Ltd., London. 1962.

6.6 Concrete Bridge Development Group (CBDG). *Notes for guidance on the assessment of concrete bridges.* Technical Guide No 9, CBDG, Camberley, UK. 2006.

6.7 Clark L.A. *Concrete bridge design to BS 5400.* Construction Press, London, UK. 1988.

6.8 CONTECVET IN30902I. *A validated users manual for assessing the residual service life of concrete structures affected by ASR.* A deliverable from an EC Innovation project. Available from the British Cement Association (BCA). Camberley, UK. 2000.

6.9 CONTECVET IN30902I. *A validated users manual for assessing the residual service life of concrete structures affected by corrosion.* A deliverable from an EC Innovation project. Available from the British Cement Association (BCA), Camberley, UK. 2000.

6.10 CONTECVET IN30902I. *A validated users manual for assessing the residual service life of concrete structures affected by frost action.* A deliverable from an EC Innovation project. Available from British Cement Association (BCA). Camberley, UK. 2000.

6.11 Webster M.P. *The assessment of corrosion – damaged concrete structures.* PhD thesis. University of Birmingham, UK. July, 2000.

6.12 British Cement Association (BCA). *Impact of deterioration on the safety of concrete structures.* Report to the DETR under the PIT programme. Project reference CI 38/13/20 (cc 1030), BCA, Camberley, UK. April, 1998.

6.13 Clark L.A. *Assessment of concrete bridges with ASR.* Bridge Management 2. Thomas Telford Ltd, London, UK. 1993. pp. 19–28.

6.14 Mulenga D., Robery P. and Baldwin R. *Multi-storey car parks – have we parked the issue of effective maintenance? Concrete Engineering International.* Vol. 10. No. 3. The Concrete Society, Camberley, UK. Autumn, 2006.

6.15 Chana P.S. and Korobokis E. *The structural performance of reinforced concrete affected by alkali–silica reaction.* Phase II Transport Research Laboratory Contractor Report 233. TRL, Crowthorne, UK. 1991.

6.16 Clark L.A. *Critical review of the structural implications of the alkali–silica reaction in concrete.* Transport Research Laboratory Contractor Report 169. TRL, Crowthorne, UK. 1989.

6.17 Narayanan R.S. and Goodchild C.H. *Concise Eurocode 2.* ISBN 1-904818-35-8. The Concrete Centre, Camberley, UK. June 2006.

6.18 Leonhardt F. and Walther R. *The Stuttgart shear tests*. Cement and Concrete Association (C&CA) Translation CJ 111. 1964. p. 134.

6.19 Regan P.E. Research on shear; a benefit to humanity or a waste of time. *The Structural Engineer*, Vol. 71, No. 19. 5 October 1993. pp. 337–347. Institution of Structural Engineers, London, UK.

6.20 British Standards Institution (BSI) *Structural use of concrete: Part 1: Code of Practice for design and construction*. BS 8110 Part 1: 1997. BSI, London, UK.

6.21 Taylor H.P.J. The fundamental behaviour of reinforced concrete beams in bending and shear. Shear in reinforced concrete. ACI SP-42. American Concrete Institute, Detroit, USA. 1974. pp. 43–77.

6.22 Chana P.S. Analytical and experimental studies of shear failures in reinforced concrete beams. *Proceedings of the Institution of Civil Engineers (ICE) Part 2, 85*. December 1988. pp. 609–628. ICE, London, UK.

6.23 Regan P.E. Shear. Concrete practice sheet no. 103. *Concrete*. November 1985. The Concrete Society, Camberley, UK.

6.24 Concrete Bridge Development Group (CBDG). *Notes for guidance on the assessment of concrete bridges*. Technical Guide No. 9. 2006. CBDG, Camberley, UK.

6.25 Collins M.P. *Reinforced and prestressed concrete structures*. Brunner-Routledge. 2nd Edition, 2005. ISBN 0419249206.

6.26 Bentz E.C., Vecchio F.J. and Collins M.P. Simplified modified compression field theory for calculating shear strength of reinforced concrete elements. *ACI Structural Journal*, Vol. 103, No 4. July–August, 2006. pp. 614–624. ACI, Detroit, USA.

6.27 Chana P.S. and Korobokis G. *The structural performance of reinforced concrete affected by alkali–silica reaction*. Phase II. Transport Research Laboratory (TRL) Contractor Report 311. 1992. TRL, Crowthorne, UK.

6.28 Cope R.J. and Slade L. The shear capacity of reinforced concrete members subjected to alkali–silica reaction. *Structural Engineering Review* No. 2. 1990. pp. 105–112.

6.29 Daly A.F. *Effects of accelerated corrosion on the shear behaviour of small scale beams*. Research Report PR/CE/97/95. 1995. Transport Research Laboratory (TRL). Crowthorne, UK.

6.30 Shave J.D., Ibell T.J. and Denton S.R. Shear assessment of concrete bridges with poorly anchored reinforcement. *Proceedings of Structural Faults and Repair*, London 2003.

6.31 Canadian Standards Association. *Design of Concrete structures*. Structures (Design) A23.3. December 1994. p. 199.

6.32 Cairns J. Strength in shear of concrete beams with exposed reinforcement. *Proceedings of the Institution of Civil Engineers, Structures and Buildings*. Vol. 110. May 1995. pp. 176–185.

6.33 Ng K.E. and Clark L.A. Punching tests on slabs with alkali–silica reaction. *The Structural Engineer*, Vol. 70, No. 14. July 1992. pp. 247–252. IStructE, London, UK.

6.34 Reynolds G.C. *Bond strength of deformed bars in tension*. Technical Report 42.548. 1982. Cement and Concrete Association, Wexham Springs, UK. (Now located in the archives of the Concrete Information Service, Concrete Society, Camberley, UK.)

6.35 Tepfers R. Cracking of concrete cover along anchored deformed reinforcing bars. *Magazine of Concrete Research (MCR)*. Thomas Telford Ltd, London, UK. Vol. 31, No. 106, March 1979. pp. 3–12.

6.36 Saifullah M. *Effect of reinforcement corrosion on bond strength in reinforced concrete*. PhD thesis. University of Birmingham, UK. April 1994. p. 258.

Chapter 7

Protection, prevention, repair, renovation and upgrading

7.1 Introduction

The need to repair concrete structures is not new. Much of the early work involved making good via patch repairs and crack filling, for aesthetic and serviceability reasons [7.1]. As the concrete infrastructure of the mid-20th century matured, there was also a demand to strengthen or upgrade to meet changes in use or increases in loadings. The need to treat cases of corrosion emerged in the 1950s with post-war prefabricated reinforced concrete housing, and many of the references to Chapter 2 detail examples of corrosion in highway structures as the use of de-icing salts increased rapidly in the early 1960s. Reference [7.2] gives some details of this, and reference [7.3] is a detailed review of the situation in the UK and France with regard to post-tensioned concrete bridges.

As durability concerns became more widespread, and the consequences of failure more critical, repair became a growth industry, and the options available on the market increased significantly in terms of principles and approaches, and the individual solutions within each basic approach. This process continues apace as witnessed via a recent publication [7.14] containing over 200 short papers on all aspects of the problem.

The literature is full of individual case studies, describing what has been physically done and giving some reasons for selecting a particular option; it is often difficult to draw general conclusions from these. Such articles, which are also helpful since they provide website addresses, appear most frequently in concrete-related journals such as *Concrete* from the Concrete Society in the UK. In North America, the various journals of the American Concrete Institute (ACI) do a similar job, and focus on repair is provided by the International Concrete Repair Institute (ICRI), which publishes a bi-monthly Bulletin, and whose website (*www.icri.org*) gives details of available publications in the USA; generally, these are either guidance documents, or compilations of articles on particular topics.

There are also guidance documents available on individual repair, protection and upgrading methods, which explain the principles involved and are strong on the 'how to...' aspects of the problem. Some examples of these can be obtained from the ICRI website for North America, and references [7.4–7.9] are similar publications available from the Concrete Society in the UK. The Concrete Society portfolio is augmented by other reports on test methods and diagnosis, and on how to enhance durability in new construction; Technical Report 61 [7.10] is an example of the latter, where much of the detailed information is transferable to the repair and renovation situation. The Concrete Repair Association in the UK also has a website (*www.cra.org.uk*).

The above brief review is intended to show that there is quite a lot of information available on repair and renovation methods and also to indicate the nature of that information. It can become dated quite quickly however, as the technology is improved and new techniques are introduced. Moreover, the nature and format of the information make it difficult to compare the technical and economic merits of alternative approaches – essential information to the owner when making a choice. This situation is now changing, with serious attempts being made to develop a systematic scientific basis for classifying repair and renovation methods, supported by sound specifications and test methods. The emergence of EN 1504 is a prime example of that, and will be referred to strongly in later sections of this chapter.

The final major missing link from the data bases is the lack of in-depth feedback on real performance in the field over relevant periods of time. How does this compare with claims and expectations? Again, this is changing, as typified by Figures 2.13–2.16, taken from the paper by Tilly [7.11]. Tilly's paper comes from the activities of a European network CONREPNET, which has examined well over 100 case studies in some detail and, apart from providing field data, has focused on developing criteria to permit alternative options to be evaluated to a common base. This information will also be used extensively later in this chapter.

Repair and renovation is a huge subject, deserving several books in its own right. This book is about assessment, management and maintenance, and repair is an integral part of that. The emphasis in this chapter is on how it fits into the overall scheme of things, in moving forward from the assessment phase to taking effective action in selecting optimum solutions. This approach leads to the following sequence of sub-sections.

7.2   Performance requirements for repaired structures
7.3   Classification of protection, repair, renovation and upgrading options
7.4   Performance requirements for repair and remedial measures
7.5   Engineering specifications
7.6   Moving towards the selection process
7.7   Performance of repairs in service
7.8   Timing of an intervention
7.9   Selecting a repair option–general

7.10 The role of EN 1504 in selection
7.11 Selecting a repair option in practice
7.12 Concluding remarks Appendix 7.1 and Appendix 7.2 References

7.2 Performance requirements for repaired structures

In simple terms, the performance requirements for repaired structures are no different from those for new construction. Structurally, the focus will be on the factors listed in Table 4.12. Progressive assessment will have led to a performance–time graph, such as that in Figure 3.13, for all relevant Table 4.12 factors. This paints a picture of how the present condition relates both to the performance levels provided in the original design and to the owner's perception of what constitutes minimum acceptable performance, bearing in mind that much more is now known about the structure (the Table 6.2 issue).

Complicating the situation is the fact that different owners may wish to manage the rehabilitation process differently. Figure 3.3 shows two viable options emanating from the asset management procedures associated with bridges in the UK. The different strategies involved intervention on different timescales, and, most probably, different solutions. Some owners may also wish to take a conservative approach, involving early preventative measures. There are no definitive general rules here, but a need to be aware of what the options are, linked to confidence in their effectiveness.

In moving forward, however, it is essential to be clear about the required performance levels. While the basic structural factors in Table 4.12 will remain, there are broader strategic issues involved, some non-technical, which will influence the course which individual owners may choose to follow. Different owners will have different strategic goals, depending, for example, on:

- type of ownership – whether private or public sector;
- changing statutory requirements;
- the type of structure and its function;
- future plans for the structure, independent of its current physical state, due, say, to

  - a possible change in use;
  - improved performance requirements arising from higher user expectations;
  - increases in imposed loadings;

- a greater emphasis on whole life costing, linked to budgetary plans;
- a desire for improved sustainability.

In a follow-up project to CONTECVET, a group of partners containing a high proportion of owners from Spain, Sweden and the UK, set out to establish a strategy for the maintenance and rehabilitation of concrete structures. As part of this project, acronym REHABCON, a list of general performance requirements was developed. Table 7.1, taken from a REHABCON deliverable [7.12], gives details. While the majority of the requirements relate to the structure as a whole, some also relate to the selected rehabilitation option and to the renovation process itself.

Table 7.1 General performance requirements for rehabilitated structures. Rehabcon [7.12]

	General performance requirements
Structural safety	Ultimate limit state design (same expectations as for new structures) • Strength (load bearing capacity, etc.) • Stability • Robustness (resistance to impact, etc.) • Fatigue • Fire resistance • Earthquake resistance
Serviceability	Serviceability limit state design (same expectations as for new structures) • Deformation • Displacement • Vibrations • Watertightness (important, for example, for dams, swimming pools, tanks) • Slip resistance/roughness (of floors, roads) • Drainage • Visibility during inclement weather (e.g. water spray on motorways) • Comfort/convenience to users (for example, limited vibrations, no dropping water on users under a structure, etc.)
Operation and function	• Availability, functionability (in working order for the intended function, for example to carry vehicles) • Minimisation of downtime. While this is important for a rehabilitated structure, it is also important to minimise inconvenience to users during the rehabilitation action, i.e. low impact on users during operation, maintenance and repair. This could be achieved by re-functioning the structure during repair, conducting quick repair works, etc. • Inspectability (it must be possible to inspect the rehabilitated structure)

Aesthetics	• Colour (e.g. the rehabilitated part of the structure shall look the same and have the same colour as the rest of the structure)
	• Texture of surface (e.g. the rehabilitated part of the structure shall look the same, have the same texture, as the rest of the structure)
	• Durability of aesthetics (the durability of the aesthetics of the rehabilitated part of the structure shall be as long as the durability of the rehabilitation action itself, or as long as the rest of the structure if this period is shorter)
	• Safe-looking (e.g. the rehabilitated structure must look safe to satisfy users)
Sustainability and environmental factors	• Materials for rehabilitation works to be sustainable, and environmentally friendly during
	• Manufacture
	• Construction works
	• Use
	• Damage (e.g. under fire conditions)
	• Demolition
	• Impact on recycling and reuse
	• Deposition (deposition of materials that can not be recycled or re-used, at least not immediately)
	• Acoustics, noise control (low noise, during rehabilitation and afterwards)
	• Energy consumption (reduce the energy consumption for rehabilitated structure, and also to consider energy consumption during all phases above)
	• Harmful effects, such as spillage, leakage, dust or the emission of toxic fumes, either spontaneously or due to situations such as fire, both during the rehabilitation works and afterwards
Health and Safety	• Public safety (e.g. barriers, railings and prevention of pieces of concrete falling from the structure due to spalling or other deterioration)
	• Health for humans and nature during all phases in the life-cycle (Safety and health for workers, users and public during rehabilitation actions etc. To consider specific safety measures to reduce risks.)
	• Evacuation, emergency escape routes
Durability	• Durability of the original structure and the rehabilitated parts of the structure. Service life [the expected durability of the rehabilitation action depends on residual life of the structure and life-cycle cost (including user costs) for the rehabilitation action]
Dependability	• Reliability of the repair methods (reduce the risk of an unsuccessful rehabilitation method)
	• Maintainability (for the rehabilitated structure
	• Maintenance supportability (for the rehabilitated structure))

Table 7.1 (Continued)

	General performance requirements
Flexibility	• Ensure that it is possible to meet future requirements (e.g. normally the rehabilitation method should not preclude future modification such as strengthening.)

Economy	• Reduce or limit whole life costs (owners cost and user cost for the rehabilitated structure)
	
	    • Operational costs
	    • Maintenance, repair and rehabilitation costs
	    • Improvements/strengthening costs
	    • Demolition and deposition costs
	    • User costs
	
	• Limit loss of income due to insufficient functionality etc. (If some of the other requirements has impact on the whole life costs of the rehabilitated structure, this cost shall be considered in the whole life cost as well (during rehabilitation and afterwards)).

Cultural Heritage	• Structures having cultural or historic value require special treatment.

The REHABCON partners recognised that Table 7.1 was a general 'catch all' and that some of the factors would be more important than others, with a different ranking system for different types of structure. Nevertheless, there were certain key requirements which could not be compromised. This led to Table 7.2, containing general key requirements and specifics for different types of structures. Table 7.2 is closely related to Table 4.12.

Table 7.2 Key performance requirements for different types of structure. Rehabcon [7.12]

(a) General requirements

Structural safety	• Strength • Stability • Robustness
Serviceability	• Deformation • Displacement
Operation and function	• Function
Sustainability and environmental factors	• Harmful effects, such as spillage, leakage, dust or the emission of toxic fumes
Health and Safety	• Public safety • Health for humans and nature during all phases in the life-cycle

(b) Additional requirements for bridges	
Structural safety	• Fatigue
Serviceability	• Vibrations

(c) Additional requirements for buildings	
Structural safety	• Fire resistance
Serviceability	• Vibrations
Sustainability and environmental factors	• Acoustics, noise control
Health and Safety	• Evacuation, emergency escape routes

(d) Additional requirements for parking structures	
Structural safety	• Fire resistance
Serviceability	• Vibrations
Health and Safety	• Evacuation, emergency escape routes

(e) Additional requirement for water-retaining structures	
Structural safety	• Water-tightness

The reality of this puts a strong emphasis on structural safety, and it can be argued that safety should be considered separately from all other aspects. Partly, this is because of its importance, but also because the solutions for raising safety levels fall into a distinct category within the full spectrum of repair, protections, renovations and upgrading options. Further, the provision of improved safety options requires special consideration of how it is actually done; structural repairs frequently involve temporary reductions in overall strength and stability, until the 'new' improvements are actually in place.

7.3   Classification of protection, repair, renovation and upgrading options

The range of available options is very large, with different groupings providing quite different potential benefits. To bring some order to this, as an aid to the selection process, it is necessary to classify these, in terms of what they do. This is important in itself, but also a major factor in the timing of any intervention. In tackling this, it is proposed to adopt the principles in EN 1504 [7.13]. There are ten parts to EN 1504, as listed in Table 7.3.

Table 7.3 Subject matter of the ten parts of EN 1504 [7.13]

General Reference	Title	Published
1504-1	General scope and definitions	Revised 2005
1504-2	Surface protection systems	October, 2004
1504-3	Structural and non-structural repair	February, 2006
1504-4	Structural bonding	November, 2004
1504-5	Concrete injection	December, 2004
1504-6	Grouting to anchor reinforcement or to fill external voids	September, 2006
1504-7	Reinforcement corrosion protection	September, 2006
1504-8	Quality control and evaluation of conformity	November, 2004
1504-9	General principles for the use of products and systems	1997 (under revision)
1504-10	Site application of products and systems and quality control of the works	December, 2003

Part 1 first appeared in 1998, and has since been published as a full EN. Text of the other Parts have also been published and are currently (2006) at various stages of moving towards formal adoption. Part 9, a key document, first appeared in 1997, and a revised version is expected in the near future.

Parts 2 to 7 detail the performance test requirements for the repair methods given in Part 9. Supporting test methods are given in numerous other CEN and ISO Standards. It is important to realise that the tests relate to repair systems but not to the performance of the repair and the structure as a system. Owners need to consider issues that could arise from the interaction between the two. An important aspect of that is the consideration of actual loads, and, especially, of the environment and local micro-climate. Part 8 covers quality control and product conformity based on test results. Part 10 deals with all aspects of site application, including the key issues of substrate condition and pre-preparation.

This leaves Part 9, the centre piece of the whole system. Its classification of options is based on 11 Principles, as set out in Table 7.4. The principles are expressed in terms of basic objectives (protection, restoration, strengthening, etc.). The methods for each objective can vary considerably, in terms of:

(i) action/function in meeting the objective (e.g. compare 1b and 1d; 4a with 4c or 4d; etc.)
(ii) method of application (e.g. compare 3a, 3b and 3c)
(iii) dependency on a product and workmanship, compared with a self-contained process (compare 7a or 7b with 7c or 7d).

Table 7.4 Principles for repair and remedial action, as given in Part 9 of EN 1504 [7.13]

1.   Protection against ingress of adverse agents (PI)
       (a)   impregnation
       (b)   surface coatings, with and without crack bridging ability
 *   (c)   locally bandaged cracks
       (d)   filling cracks
 *   (e)   transferring cracks into joints
 *   (f)   erecting external panels
 *   (g)   applying membranes

2.   Moisture control (MC)
       (a)   hydrophobic impregnation
       (b)   surface coatings (A: sealing, or B: paint)
 *   (c)   sheltering or overcladding
 *   (d)   electrochemical treatment

3.   Concrete restoration (CR)
       (a)   applying mortar by hand
       (b)   recasting with concrete
       (c)   spraying concrete or mortar
 *   (d)   replacing elements

4.   Structural strengthening (SS)
       (a)   adding or replacing embedded or external reinforcement
       (b)   installing bonded rebars in preformed or drilled holes in the concrete
       (c)   plate bonding
       (d)   adding mortar or concrete
       (e)   injecting cracks, voids or interstices
       (f)   filling cracks, voids or interstices
 *   (g)   prestressing – post tensioning with external cables

5.   Physical resistance (PR)
       (a)   overlays or coatings (A: wearing surface, or B: membrane)
       (b)   impregnation

6.   Resistance to chemicals (RC)
       (a)   overlays or coatings (A: wearing surface, or B: membrane)
       (b)   impregnation

7.   Preserving or restoring passivity (RP)
       (a)   increasing cover to rebars, with additional cementitious mortar or concrete
       (b)   replacing contaminated or carbonated concrete
 *   (c)   electrochemical re-alkalisation of carbonated concrete
       (d)   re-alkalisation of carbonated concrete by diffusion
 *   (e)   electro-chemical chloride extraction

8.   Increasing resistivity (IR)
       (a)   limiting moisture content by surface treatment, coatings, sheltering, or hydrophobic impregnation

9.   Cathodic control (CC)
       (a)   limiting oxygen content at the cathode, by saturation or surface treatment

Table 7.4 (Continued)

10.	Cathodic protection (CP)
	(a) applying electrical potential (A: passive, or B: active)
11.	Control of anodic areas (CA)
	(a) painting reinforcement with coatings containing active pigments
	(b) painting reinforcement with barrier coatings
*	(c) applying inhibitors to the concrete

Note: Those marked* may involve products or systems which are outside the scope of the EN 1504 series (Table 7.3).

On the other hand, some of the methods appear against more than one principle (e.g. 1b, 2b and 8a, for surface coatings), possibly requiring different formulations to meet the different objectives.

For these reasons and because performance criteria have to be considered in the context of engineering specifications, it is necessary to re-order the various methodologies to reflect better the physical or chemical actions involved, and hence bring out the essential differences in engineering specifications. Further, Table 7.4 has a strong emphasis on repair and remedial work where the principal deterioration mechanism is corrosion; other forms of deterioration (e.g. ASR, frost action, or others which directly attack the concrete) may require a more limited set of repair options, which can be fitted into a re-ordering of Table 7.4.

Table 7.5 presents the outcome of this re-ordering, to facilitate the definition of performance requirements and specifications. In general, the five categories in the table are presented in the order of decreasing direct structural intervention, with category 1 being the most significant and category 5 the least.

The timing for introducing a chosen repair or remedial measure will depend on the owner's strategy and on the criteria selected for decision-making. For example, with corrosion, the owner may wish to intervene at any of the following stages of deterioration:

(i) before carbonation or chloride fronts have reached the reinforcement;
(ii) when the fronts have just reached the reinforcement;
(iii) when corrosion has just started;
(vi) when corrosion has reached a certain maximum depth;
(v) when cracking or spalling of the concrete cover is about to occur;
(vi) when a maximum allowable loss of rebar section has occurred.

The selection of an option from Table 7.5 will plainly depend on the nature and extent of the observed damage. It will also depend on the extent of the inspection and investigation (associated with confidence levels) and on the sensitivity of the structure as well as the owner's future plans for that

Table 7.5 Re-classification of repair options, according to physical function or method

Option	Comments
1. External strengthening or barriers (a) replacing complete elements (b) plate bonding (c) external prestressing (d) provision of additional concrete cover (with or without additional rebars)	(a) is used, when deterioration is extensive, and disruption to operations is minimal. (b) and (c) are used where additional strength is required. Calculations will be required. For (b), bond and adhesion are important, and it is necessary to demonstrate composite action, while ensuring compatibility of materials. (d) may also be used when additional strength is required, or to provide extra protection against the risk of corrosion. Method of installation is important in ensuring good bond and an effective dense concrete.
(e) waterproof membranes or sheets (f) enclosures or physical barriers	(e) is common practice for bridge decks and specifications do exist. Enclosures have also been used for bridges, to control local micro-climates. Generally, physical barriers may be either separate or bonded to the parent structure; examples are cladding to bridge sub-structures.
2. Structural repairs (a) cutting out contaminated, cracked or defective concrete and replacing it (b) cutting out and replacing corroded reinforcement (c) cutting out concrete, adding protection to the reinforcement and replacing the concrete	Variations of these methods are aimed at restoring or increasing strength, within the perimeter of the existing structural elements. (a) may be used when the damage has been caused by chemical or physical actions, such as ASR, frost action or abrasion. (b) and (c) relate to corrosion damage. These are structural repairs, and the temporary conditions during repair operations require structural consideration, with adequate propping being provided where necessary. The materials used and their formulations, may depend on the size of the areas to be replaced. Many proprietary systems are on the market, and it is important to ensure compatibility with the parent concrete, as well as full composite action. Pre-preparation is crucial, to ensure bond both with the substrate and the rebars, and good workmanship is paramount, for all the application methods which may be used.

Table 7.5 (Continued)

Option	Comments
3. Surface coatings and impregnations Surface treatments Coatings Impregnations	Essentially, the objective here is to prevent the ingress of adverse agents, or to control moisture (but see the different principles in Table 7.4). There are numerous proprietary systems on the market, with a range of formulations. It is important to understand the nature of each, exactly how it operates and what is necessary for effective installation. The likely life of each should be established within reasonable bounds, for the particular location and environment. Surface treatment may involve making good minor defects, followed by a coating. Surface coatings, to be effective, require a smooth surface and to have the ability to accommodate movement (including cracks) without fracture to the protective film or barrier. Impregnations, by definition, must be able to penetrate the concrete to a minimum specified depth, commensurate with their effectiveness.
4. Filling cracks and voids	The materials or techniques used depend on the size of the voids or the width of the cracks. Surface voids or defects can be treated as patch repairs. For the effective filling of cracks, injection is often used. Again, there are many systems on the market and choice is important, while ensuring that the materials used comply with relevant product standards. This technique is often complemented by a surface coating or treatment.
5. Electro-chemical techniques (Principles 7, 9, 10 and 11 in Table 7.4) (a) re-alkalisation (b) chloride extraction (c) cathodic protection (d) cathode or anode control	These techniques are all grouped together because they are effectively self-contained processes, which require Performance Standards – unlike categories 2–4 above, which relate more to product Standards supported by acceptable test methods and backed by specification for installation and workmanship. All have been used with some success, with cathodic protection perhaps the most common and already having its own specification/standard.

structure. For corrosion, categories 3–5 in Table 7.5 are possibly most relevant for stages (i)–(iii) above, with categories 1 and 2 coming into play for stages (iv)–(vi). Whatever remedial technique is selected, the timing of the intervention will influence both the specification and the performance requirements for that technique.

The primary role of the options listed in Table 7.5 is to maintain or restore the performance requirements in Table 7.2 to an acceptable level. As may be seen, there are numerous ways of doing that; whichever option is selected, its performance requirements must be clearly established, to ensure the required structural performance. However, the choice of the method of implementing the principle will be influenced also by the Health and Safety regulations in existence (Table 7.1).

Moving from Table 7.4 to Table 7.5 is intended to make it easier to define performance requirements for the different options, without changing the basic principles in EN 1504 – and to integrate these into engineering specifications. In using the EN 1504 system, the basic performance requirements are given in Parts 2–8, which, in turn, will call up other CEN Standards on test methods, compliance, etc. In the context of this book, only a flavour of this procedure can be provided, and this is done in Sections 7.4 and 7.5, using the engineering approach in Table 7.5 as a base.

7.4 Performance requirements for repair and remedial measures

7.4.1 Introduction

Performance requirements, in an engineering context, depend on the nature of the repair or remedial measure and on what is to be achieved by using it. Referring to category 1 in Table 7.5, methods (a), (b) and (c) are concerned with providing additional strength. This may also be a role for method (d), associated with providing extra protection against the ingress of contaminants.

Methods (e) and (f) are classed as external means of providing physical barriers to control moisture movements or exclude contaminants – although the action of each may be very different in doing so.

Category 2 in Table 7.5 – structural repairs – has been isolated out, because it is again concerned with enhancing strength and serviceability – this time internally, by modifying the elements in the structure itself. The structural aspect is emphasised because, in effecting the repair, it will be necessary to consider the temporary load-bearing function, as well as adequacy in service, during repair operations.

While the methods listed in categories (3)–(5) in Table 7.5 vary very much in nature and function, they do not directly affect the load-bearing capacity of structural elements by their action. The prime objective is to prevent (or slow down) future deterioration or to restore (or enhance) the resistance of the structure to contaminants. In some methods intrusion into the concrete is necessary, but this is usually minor in terms of possible influence on structural capacity.

A distinction is therefore being made between methods where structural capacity is affected in some way and methods targeted more at preventative measures. This difference is significant, in terms of engineering performance requirements, both during the repair and subsequently on service.

The performance requirements in support of the Table 7.4 principles are contained in Parts 2–8 of EN 1504. These are still evolving, and being added to, as new innovative options appear.

A brief engineering perspective is provided in Sections 7.4.2–7.4.5 for each of the five categories of options in Table 7.5.

7.4.2   External strengthening of barriers (category I in Table 7.5)

The six methods in this category have the common feature that they all involve adding some external item to the structure. However, they do differ in terms of function and objective and hence in terms of performance requirements.

7.4.2.1   Element replacement

This involves the removal of a badly deteriorated element and its replacement by something equivalent. The new element will be designed to current design standards and this sets both the performance requirements and specification.

Removal of an element can be difficult in practical terms. Moreover, it is also crucial to consider the strength and stability of the structure when the element has been removed, in assessing load-carrying capacity. The remaining engineering requirement is to join the new element to the present structure, to achieve the required degree of determinacy.

7.4.2.2   Plate bonding

Additional strength may be provided by bonding either steel or non-ferrous plates, usually to beam or slab elements. Non-ferrous materials have also been used to bind and strengthen sub-structures such as circular columns.

The key engineering feature is the need to ensure composite action between the plate and the parent element. This requires good adhesion and bond characteristics, and good surface preparation to ensure that the selected bonding agent performs to the specification.

7.4.2.3 External prestressing

This technique is most suited to major structures such as box-beam bridges, where geometry permits the installation and protection of external post-tensioning tendons. To be effective, the external prestressing force has to be transmitted to the parent structure. This requires the installation of anchorages and end blocks, and of intermediate diaphragms or saddles. The design and detailing of these have to be done with some care so that the structural integrity of the parent element is not affected. This is a specialised technique generally carried out by experienced prestressing companies.

7.4.2.4 Provision of additional concrete cover

This technique has been used for the top surfaces of bridge decks and also for bridge soffits, usually by spraying. It has also been used in tunnels and buildings where additional fire protection is required or where concrete cover is low.

To be effective, good bond is essential, which again requires care in surface preparation. The formulations of the materials used are generally cement-based and selection involves consideration of how to avoid cracking due to either shrinkage or differential movement between the new and parent concretes; for significant thicknesses, mesh reinforcement may be required.

7.4.2.5 Waterproofing membranes

The use of surface coatings or impregnations for moisture control is dealt with in category 3. This sub-section refers to the use of physical membranes, as used typically on the top surface of bridge decks, bonded to the concrete and located beneath the surfacing.

The performance requirement is for an effective continuous barrier, largely unaffected by traffic loading. Specifications do exist for this application, with a strong emphasis on jointing and on the detailing of edges and boundaries.

7.4.2.6 Enclosures or physical barriers

These are defined as external physical means of creating barriers to the ingress of contaminants or for the control of local micro-climates, especially moisture movements. The exact performance requirements need clear definitions in individual cases, but essentially the objective is the reduction or elimination of aggressivity in the local micro-climate.

Enclosures, usually made of non-degradable materials, have been used to protect bridge soffits from the effects of spray due to de-icing salts; this is a specialist technique but guidance documents on performance requirements do exist. More generally, for physical barriers, the norm is to bond them to the parent structure, e.g. overlays on horizontal surfaces, or cladding for vertical sub-structures.

7.4.3 Structural repairs (category 2 in Table 7.5)

All three methods listed in Table 7.5 are variations of the same basic operation which, depending on the deterioration mechanism and the extent of the damage, involves cutting out and replacing concrete and/or reinforcement. The operations are within the perimeter of the existing structural elements and the primary structural performance requirement is to restore or enhance the original strength, stiffness and serviceability.

It is important to remember that these are structural repairs and, therefore, it is necessary to consider structural issues:

- under the temporary conditions during the repair operations, and
- for the long-term in-service performance of the repaired elements.

To meet the structural performance requirement, the object is to ensure that the repaired element acts as a composite entity. To achieve that, a number of issues have to be considered as itemised below.

7.4.3.1 Pre-preparation

It is important to identify the cause of the deterioration and the extent of the degradation or contamination; the objectives are to restore structural capacity and to minimise the risk of the deterioration occurring again. An appreciation of the nature of the repair (material and installation) is also necessary; this involves ensuring that the material complies with the relevant product standard and that method statements exist, compatible with the selected placement method and the necessary standard of workmanship.

Pre-preparation is also important regarding the size and perimeter of the area to be repaired. This includes:

- cutting out to sound uncontaminated or underteriorated concrete, using methods which do not damage the substrate;
- making sufficient space to permit bonding (or lapping) of reinforcement;

- surface preparation of the concrete, in terms of cleaning, removing of loose material and creating conditions for effective bond and composite action;
- cleaning reinforcement and applying protective coatings when required;
- avoiding feathered edges around the perimeter.

It is also necessary to ensure easy access and that the temporary props are properly installed and pre-loaded as necessary. Provision for storage of materials and the keeping of records of all operations are also required.

7.4.3.2 Basic compatibility between repair material and parent concrete

The repair material should be checked for the following requirements:

Strength	Life time expectancy
Stiffness	Thermal response
Density	Moisture movement
Permeability	Shrinkage and creep
Fire resistance	Visual appearance
uv resistance	

7.4.3.3 Compatibility, with respect to effectiveness of repair

The following require consideration:

Bond	Restrained shrinkage/expansion
Curing requirements	Consistency, including quality assurance

Placeability (i.e. workability and setting time, commensurate with installation method).

7.4.3.4 Performance requirements in special circumstances

Depending on the environment and location of the structure and its function, consideration may have to be given to additional performance requirements. These might include:

Chemical resistance	Frost resistance
Abrasion	Leaching
Hardness	Water-tightness
Slip-resistance	Effectiveness under dynamic or fatigue loading

7.4.3.5 Health and safety

Repair materials should be non-toxic and comply with all health and safety legislation.

7.4.4 Creating barriers to contaminants and/or moisture control (categories 3 and 4 in Table 7.5)

In terms of performance requirements, categories 3 and 4 in Table 7.5 are considered together, since the basic objectives are broadly the same, differing only in terms of how these are met and in the characteristics necessary to ensure their effectiveness. The basic objective is to create a barrier to the ingress of contaminants or to control moisture movement. In functional terms, coatings are required to provide a thin but continuous barrier at the surface, impregnations require penetration of the concrete either to block the pore system or to change the resistance of the concrete to the passage of water and crack fillers may have to penetrate deeper into the concrete (and hence have a greater need for material compatibility).

The performance requirements for the wide range of materials on the market are treated in some depth in EN 1504 and its supporting standards and only a brief outline of the more important points are given here.

7.4.4.1 General requirements

As for structural repairs, the creation of barriers is sensitive to the quality of the pre-preparation and to variations in standards of workmanship during installation. Much background data has been obtained under laboratory conditions and the relevance of this to particular site conditions requires careful evaluation. While compliance with relevant product standards should be demonstrated, this has to be accompanied by method statements for site application, compatible with the installation method and the necessary standards of workmanship.

The objective of the barrier has to be clearly established (exclusion of contaminants and/or moisture control) for the particular micro-climate, and its life-time expectancy evaluated. Effective adhesion is a performance requirement which is common to all types of barrier.

In addition to its primary functions, a barrier may have to be resistant to other types of action which may be present depending on the location and function of the structure. A list of some possibilities is given in Section 7.4.3.

There are many different products on the market, from a wide range of generic types. All materials should be non-toxic and should comply with health and safety legislation.

7.4.4.2 Coatings

To be effective, coatings must adhere to the concrete surface and be continuous. This creates a list of performance requirements which should be checked; this includes:

Adhesion
Resistance to chlorides and CO_2
Effective thickness
Sensitivity to the local environment (uv-radiation; weathering; etc.)
Crack bridging ability
Resistance to moisture and water vapour
Thermal resistance
Extensibility/plasticity (to be able not to crack under strain)

The range of generic types of coatings on the market may give different relative performances for each of these; selection is therefore of great importance in individual applications, in matching required performance to desired life-time expectancy.

7.4.4.3 Impregnations

To be effective, impregnations have to penetrate to a minimum depth and it is important to check that this is feasible, for the quality of the particular concrete. Alkali-resistance should also be checked.

Impregnations act either by blocking the pore system of the concrete, or by changing its resistance to the passage of water (hydrophobic). Consideration, therefore, has to be given to the different transport mechanisms (e.g. permeability, diffusion, capillary absorption, etc.) for moisture and water vapour, in making a selection.

7.4.4.4 Crack fillers

Performance requirements for crack fillers will depend on the width and depth of the voids and on whether or not the cracks are live or dead. Cracks may be filled for purely aesthetic reasons but, more commonly, to improve the resistance of embedded metal to corrosion caused by chlorides or carbonation. This immediately define the key performance requirements.

In addition to the above general requirements some additional requirements should be checked. These include:

Volumetric change Compatibility with the parent concrete
Viscosity   Injectibility, where appropriate
Water-tightness

7.4.5 Electro-chemical techniques (category 5 in Table 7.5)

In pure performance terms, there is little that can be said at a detailed level about these techniques. The objective of each is implicit in the title and the intentions are to address corrosion issues by restoring passivity in some way, to eliminate or slow down future corrosion.

Mostly, these involve self-contained processes, which require the development of track records, leading to performance-based standards. Some involve permanent installations; others are relatively short-term treatments (less than 3 months). The most developed technique is cathodic production, with successful field experience leading to a CEN Standard (EN 12696), giving performance criteria for setting up such a system.

7.5 Engineering specifications

7.5.1 Introduction

A decision to intervene will be based on data from investigations and assessment. The nature of the intervention will depend on these findings and also on the management and maintenance strategy of the owner. The over-riding feature is that the structural performance should be maintained for a specified time; that period may be the remainder of the structure's useful life, or a designated lesser period, at the end of which a further assessment will be undertaken and future management and maintenance planned. Whatever it is, it should be specified. The essence of this is to have a life-time performance plan and a perception of what is acceptable in terms of minimum performance for the key factors listed in Tables 7.1 and 7.2.

Tables 7.4 and 7.5 give the principles and options on which decisions relating to the nature of the intervention can be based. In simple terms, these reduce to:

(a) direct structural solutions, in terms of repair or upgrade (categories 1 and 2 in Table 7.5);
(b) indirect solutions, in creating barriers of some sort, to eliminate or minimise the risk of further degradation due to the effects of contaminants (categories 3 and 4 in Table 7.5); and
(c) special techniques, which are performance-related but having the same fundamental objective as (b) above.

The performance requirements for these three options are given in Section 7.4. Having matched them against what is known about each solution, a decision is made to adopt a particular course of action. In taking that forward, a translation has to be made into engineering, material and process specifications for the work. The principles on which that might be done are outlined briefly in the sub-sections which follow.

7.5.2 Strengthening options (categories 1 and 2 in Table 7.5)

These are direct structural solutions either external (category 1, Table 7.5) or internal (category 2, Table 7.5). It follows that the adequacy of the solution must be demonstrated, either by direct calculation (category 1(a)) or by calculations supported by physical evidence of adequacy (virtually every other method in categories 1 and 2, except 1(c)). Where calculations are required, the methods to be used should be specified.

For solutions which involve the introduction of new materials or elements, either externally or internally, specifications are required for:

(i) the materials to be used,
(ii) the methods of preparation including standards of workmanship, and
(iii) the processes for pre-preparation and for ensuring an effective connection between the new material and the parent concrete – all to ensure composite action.

Item (i) can come from relevant product standards, supported by other standards on test methods etc. Item (ii) can come from manufacturers' literature, while ensuring that the recommendations apply to the particular location and site conditions. The pre-preparation part of item (iii) can also be based on manufacturers' literature; however, the key part of this is evidence, from realistic tests or site trials, that composite action is assured as assumed in the re-design. Items (ii) and (iii) together require a method statement to be developed and QA procedures to be in place, to ensure that this is followed.

7.5.3 Barrier solutions (categories 3 and 4 in Table 7.5)

The required performance requirements are itemised in Section 7.4.4. The solutions are product-specific, covering a wide range of different generic materials. Ideally, the material part of the specification should refer to the relevant product specification. The installation part is more difficult but equally important, since feedback from practice has shown that most solutions are sensitive to pre-preparation and to standards of workmanship. Again, method statements are essential, ideally based on research data relevant to site conditions. Manufacturers' literature is the basis for this, plus experienced personnel and adequate QA procedures in place.

7.5.4 Specialist techniques (category 5 in Table 7.5)

As far as is known, only cathodic protection has a standard, giving performance requirements and procedures for installation; plainly, this should be used for specification purposes.

Other techniques are highly process-related and the amount of experience in their use is limited, but generally positive. Such processes need to be performance-based, as with cathodic protection and, currently, specifications can only be based on the recommendations of the specialist suppliers.

7.6 Moving towards the selection process

Section 7.2 briefly covers the performance requirements for repaired structures. Section 7.3 puts forward a classification system for protection, repair and upgrading options. This is followed by performance requirements for these options (Section 7.4), together with some notes on engineering specifications (Section 7.5).

All this represents the gathering of basic information in moving towards the selection process (covered later in this chapter after a review of the feedback available on the performance of repairs in the field). The entire basis so far stems from EN 1504, and a brief look is now taken at how the EN 1504 procedures are perceived, in terms of integration into the progressive assessment process as a whole. This is illustrated in Figure 7.1, which, in effect, is a follow up to the basic action-module in Figure 4.1

The left-hand side of Figure 7.1 is the basic flow chart, with the right-hand side outlining the essential inputs at the different levels. Performance requirements for the structure appear first of all, followed by decisions on what type of option might be best and consideration of the performance requirements for the chosen option, in that order.

All this seems relatively straightforward, but is not so easy in practice. It is common for there to be more than one possible solution, which have to be judged first on their relative technical performance and on how early intervention is warranted. Economic judgements also have to be made, not only of the direct costs of the different options, but, more importantly, of the whole-life costs (life cycle cost analysis – LCCA). However done, LCCA must include an evaluation of the effectiveness of the repair over the required life, in continuing to meet the essential structure-specific requirements. This should entail a study of the available feedback on real performance – the subject covered in Section 7.7 – prior to looking at criteria for selection and reviewing alternative methods available in making a choice.

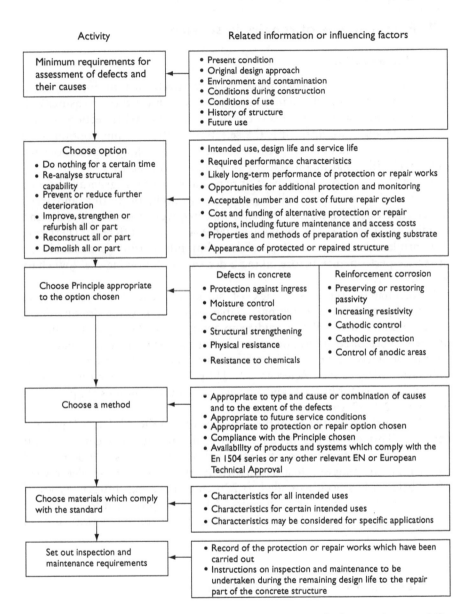

Activity

| Minimum requirements for assessment of defects and their causes |

Related information or influencing factors

- Present condition
- Original design approach
- Environment and contamination
- Conditions during construction
- Conditions of use
- History of structure
- Future use

Choose option
- Do nothing for a certain time
- Re-analyse structural capability
- Prevent or reduce further deterioration
- Improve, strengthen or refurbish all or part
- Reconstruct all or part
- Demolish all or part

- Intended use, design life and service life
- Required performance characteristics
- Likely long-term performance of protection or repair works
- Opportunities for additional protection and monitoring
- Acceptable number and cost of future repair cycles
- Cost and funding of alternative protection or repair options, including future maintenance and access costs
- Properties and methods of preparation of existing substrate
- Appearance of protected or repaired structure

Choose Principle appropriate to the option chosen

Defects in concrete	Reinforcement corrosion
• Protection against ingress	• Preserving or restoring passivity
• Moisture control	• Increasing resistivity
• Concrete restoration	• Cathodic control
• Structural strengthening	• Cathodic protection
• Physical resistance	• Control of anodic areas
• Resistance to chemicals	

Choose a method

- Appropriate to type and cause or combination of causes and to the extent of the defects
- Appropriate to future service conditions
- Appropriate to protection or repair option chosen
- Compliance with the Principle chosen
- Availability of products and systems which comply with the En 1504 series or any other relevant EN or European Technical Approval

Choose materials which comply with the standard

- Characteristics for all intended uses
- Characteristics for certain intended uses
- Characteristics may be considered for specific applications

Set out inspection and maintenance requirements

- Record of the protection or repair works which have been carried out
- Instructions on inspection and maintenance to be undertaken during the remaining design life to the repair part of the concrete structure

Figure 7.1 Overview of how the EN 1504-9 procedures might fit into the progressive assessment process (Reproduced with permission from reference 7.16. HIS, BRE Press, 2007)

7.7   Performance of repairs in service

7.7.1   Introduction

The classification of remedial measures is relatively easy to do, as indeed, is the setting out of performance requirements both for these measures and for the repaired structure. This is covered briefly in earlier sections in this chapter. Much more difficult is the selection of the optimum option for a particular case, and to attempt that rationally, in matching expectations for the repair with what might actually be achieved in practice.

The evolution and development of remedial measures is generally backed by appropriate scientific and engineering research, by testing and trials under controlled conditions. Mostly, this involves laboratory work, and conditions in service can be both variable and very different. Therefore, it is prudent to consider, if possible, what information is available on in-service performance before taking a decision in following the flow chart in Figure 7.1.

Unfortunately, there is a lack of sufficient general information on in-service performance to permit reliable conclusions to be drawn. There are papers on individual case studies involving particular repair options for different types of structure, but most do not have sufficient information to be truly definitive. Several of the references to Chapter 2, between 2.30 and 2.40, are typical of this. Reference [2.41] by Tilly is an exception, covering 70 case histories for bridges, and a brief review of the findings are presented in Figures 2.13–2.16.

Tilly's work emanated from the activities of CONREPNET, a European network. This group has produced detailed guidance, currently in draft form, on the selection process for remedial measures. This will be referred to later in this chapter. Just as important, the original in-service review of 70 bridges has been extended to 230 cases covering different types of structure in a variety of environments [7.15]. It is proposed to review the findings in some detail in Section 7.7.2, since these are of practical value in the selection process.

7.7.2   Lessons from the CONREPNET survey [7.15]

7.7.2.1   Scope

Table 7.6 provides an outline of this project, in terms of its nature and scope. A total of 230 structures were included, covering six different types of structure in 11 countries having variable environments. An indication is given of the ages of the structures where repairs were undertaken, together with the primary causes of the original deterioration. The types of repair are also itemised, together with an indication of frequency of use.

Table 7.6 Outline details of the CONREPNET survey of in-service performance of repairs [7.15]

Type and number of structure	Buildings	Bridges	Dams	Car parks	Power stations	Other	Total number
	77	75	36	8	12	22	230

Location and nature of environment	11 countries, from Scandinavia in the north to Greece and Spain in the South. Environments classified as urban (84); rural (54); highway (54); coastal (27); industrial (11)
Age factors	The majority of the structures (61%) were built between the early 1950s and the 1980s. 18% were more than 60 years old. When repaired, 80% were in the age range 0–40 years, and 50% in the range 11–30 years. 8% were less than 10 years old and the oldest was over 100 years.
Primary causes of original deterioration	Corrosion ≈ 55%; Frost action ≈ 10%; cracking generally ≈ 9%; ASR≈ 6%. Other causes, including faulty construction, leaching, scour, impact, movement ≈ 20%.
Types of repair	For the bridge component of the sample, the distribution of types of repair are shown in Figure 2.13. More generally, the total number of repairs was approx. 365 (more than one type being applied to a number of structures). The overall breakdown of repair types was:- patching 138; coatings 72; crack injection 65; strengthening 42; sprayed concrete 30; electro-chemical 18.
Notes on repair type and usage	Patching was used in 60% of the case studies, with cementitious material 60% of the time and polymer modified material for 30%. Other systems included the use of fibres. Coatings were used for 35% of the case studies, the most common being of the barrier type (60%) and hydrophobic (25%). Patching and coatings were sometimes used in combination. Strengthening generally involved replacing corroded reinforcement. Adding reinforcement and concrete also featured, as did plate bonding and added prestress.

Total number 230

This data base is considerably larger than that reported in reference 2.41, and the variety of structural types has meant that some of the trends logged in Figures 2.13–2.16 have been modified without being fundamentally changed. In attempting to get a more quantified grip on the findings, it should also be noted that the data are limited on some issues; particular examples are:

– only 6 per cent of cases involve ASR;
– the use of electro-chemical techniques is limited;
– The number of strengthening cases involving plate bonding or additional external prestress is limited.

It is also worth noting that 20 per cent of the cases did not involve primary deterioration mechanisms such as corrosion, ASR and frost action, but were due to faulty construction, imposed loads or deformations and actions such as scour or leaching.

Nevertheless it represents a valuable source of data, which gives an indication of how the reality of repairs matches expectations and clues on what are the key factors which control the differences. The key lessons from this will now be drawn out, as guides to the selection process covered later in this chapter.

7.7.2.2 Expectations for repairs with different management strategies

Safety consideration will dominate for all types of structures, where the consequences of failure are high, especially with respect to any possible loss of life. Sensitive structures, such as nuclear power plants with lives of approximately 40 years, have to be in a safe operational condition at all costs; in consequence, any repair should last to the end of that designated life.

Within different strategies and management systems, expectations for repairs can vary, depending on the type of structure and owner attitude. Two examples are:

1 *Transport structures.* Life expectancy is usually quoted as more than 100 years. The structures have to remain safe and serviceable, with minimum disruption to traffic for remedial action. Repairs might be expected to last for at least 25 years.
2 *Commercial structures.* This can cover a wide range of structural types and functions, with varying design lives, but a dominant requirement is to remain operational with minimum downtime. In the past, there has been a tendency here towards speedy repairs or preventative measures, which may not always be the best long-term technical solutions. Often, value-for-money concepts have led to the cheapest option, arrived at via competitive tendering.

This picture is now undergoing change. While the past performance of repairs may have been disappointing, the demand for more lasting and reliable solutions is on the increase. Requests for guarantees for repair systems are more common. Owners are becoming more conscious of the need for Life Cycle Analysis (LCA) in matching required performance to minimum overall cost. This is also driven by sustainability factors.

Against that background, and in the face of these trends, lessons from past performance can be of great benefit.

7.7.2.3 Classification of performances

CONREPNET used the following classification system, as at the most recent inspection.

1   *Successful*   Needing no immediate attention
2   *Evidence of Failure*   Unsatisfactory, and eventually requiring further attention
3   *Failed*   Requiring immediate attention

Using this system for all types of repair, 50 per cent were classified as successful, with the remaining 50 per cent split evenly between Evidence of Failure and Failed.

Breaking this down into types of repair, the success rates were as follows:

Patch repairs   50 per cent, with polymer modified materials being marginally better than cementitious patches
Coatings   50 per cent, with hydrophobic coatings being marginally better than barriers
Crack injection   70 per cent
Sprayed concrete   30 per cent
Strengthening   75 per cent
Electro-chemical   35 per cent

These figures are indicative, and have to be seen in context, with respect to small samples in some cases (Table 7.6). There was also clear evidence that some repair systems worked better in combination, e.g. patches plus coatings.

Of further interest from the general analysis is the time scale when failures occurred. For all cases, 20 per cent of failures occurred within 5 years, 55 per cent within 10 years, and 95 per cent within 25 years. These numbers have to be looked at closely, when considering what modes of failure were identified, and, especially, when identifying the possible causes of the failures.

7.7.2.4 Success rates in relation to deterioration mechanisms

(a) *Corrosion.*   Overall, 50 per cent of repairs to corrosion were successful, with 70 per cent success for crack injection and 40 per cent, for patches. Crack injection could be classified as a protective measure when corrosion is at an early stage. The more sensitive process of patching, involving the removal of affected concrete and the cleaning of the reinforcement, can depend on the care taken in pre-preparation, first in terms of workmanship and second, in deciding the extent of the concrete to be removed. Ensuring effective bond between the substrate and a compatible patch material is also important.

(b) *ASR.*   The number of ASR cases was low, and the success rate only 20 per cent. The reported types of repair were patching and coating, either alone or in combination. There was no sufficient evidence to indicate why the success rate was so low, except that the ASR-action may not have been complete. Coatings were presumably targeted at minimising the uptake of water, and patches, at repairing cracks or replacing any spalled concrete.

(c) *Frost action.*   Here also the success rate was low at 25 per cent, without any real indication as to what was done. Frost action does not have a high profile in the UK, but the two principal effects are well understood: surface scaling in the presence of salts and internal mechanical damage in the form of micro-cracking.

(d) *Cracking in general.*   Here, the success rate was 65 per cent, suggesting that the processes of crack filling or injection are fairly well established.

7.7.2.5 Modes of repair failure

Overall, this is shown in Figure 7.2, which is an updated version of Figure 2.15, showing a relative increase in debonding failures. The 'Other'

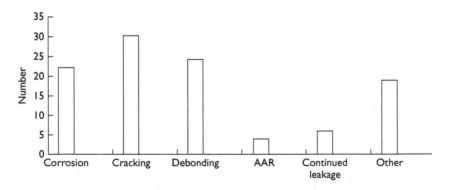

Figure 7.2 Modes of failure – all types of repair (Reproduced with permission from [7.15]. IHS BRE Press, 2007)

category included deterioration in the concrete, including some cases of spalling, plus deterioration in the coatings themselves.

For particular types of repair, the statistics were as follows:

(a) *Patches.*   Among failures, 30 per cent were classed as cracking, 25 per cent as debonding, 25 per cent as continuing corrosion and 20 per cent as other modes.
(b) *Coatings.*   Among failures, 25 per cent were classed as cracking, 25 per cent as debonding, 20 per cent as corrosion, 10 per cent as continuing ASR action, and 20 per cent as other modes.
(c) *Sprayed concrete.*   Failure modes were mainly cracking, debonding and continuing corrosion.
(d) *Cathodic protection.*   Failures were mainly due to faults in the system, e.g. with electrical connections or anode failure, which, in general, could be easily rectified. Only in one case was there failure to arrest the corrosion process.

In selecting, designing and installing these particular repair systems, it is important to note the significance of cracking and debonding as frequent failure modes. For patches, continuing corrosion may have been due to not removing sufficient concrete or a failure to appreciate the nature and location of the corrosion cells in the structure as a whole. With coatings, it should also be remembered that, in some cases, the coatings themselves deteriorated. Basically, the bulk of the failures are due to practical aspects of design and installation, once the choice has been made of a particular option, leading to the identification of attributed causes of failure in Section 7.7.2.6 below.

7.7.2.6 Causes of repair failures

It is difficult to identify a single cause for a particular failure, since, frequently, there are several reasons which are inter-related. The classification used in the CONREPNET project was:

– Incorrect design of the repair
– Use of incorrect materials
– Poor workmanship
– Wrong diagnosis
– Other factors

The breakdown into these classes is shown in Figure 7.3, an updated version of Figure 2.16.

There is a subjective element to this classification, and, in particular, the first three columns in Figure 7.3 are inter-related. Poor workmanship was a general problem, but there were instances where execution was difficult

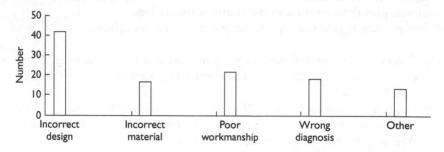

Figure 7.3 Reported causes of failure (Reproduced with permission from [7.15]. IHS BRE Press, 2007)

in practical terms, due to the choice made of a particular option involving materials and a method difficult to actually carry out in the field. Repair materials were rarely inadequate in themselves but not always compatible with the substrate in terms of placing or even mechanical properties such as strength and stiffness. Pre-preparation is crucial, as is the establishment of proper specifications and method statements. The number of cases of wrong diagnosis is perhaps surprising. The dominant deterioration mechanisms can usually be established, but more than one may be at work, and cracking can occur for a variety of reasons.

7.7.2.7 Influence of environment

The definition of different environments is given in Table 7.6. Figure 7.4 shows the success rate for all repairs in each of these.

Those located in an urban environment had a high success rate, while those in coastal and industrial areas performed poorly, possibly due to the

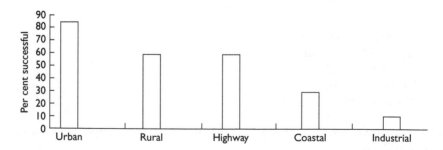

Figure 7.4 Influence of environment on performance – all types of repair (Reproduced with permission from [7.15]. IHS BRE Press, 2007)

aggressivity of the local micro-climate. General climatic conditions have to be considered for all categories of repair in Table 7.5, but moisture state and temperature are particularly important for categories 2–4, and could influence not only the choice of option but also the selection and specification of materials within that option.

7.7.3 General conclusions from the CONREPNET project

There is considerable scatter in the results from this extensive survey, in the sense that apparently identical repairs can perform very differently. This suggests that currently available methods and guides, while improving all the time, are either not being applied properly or are deficient in some respects. Clearly, the art of repair is a sensitive operation, requiring the right combination of design, materials and execution. In fact, rather like concrete construction in general.

As part of their work, the CONREPNET team surveyed a total of 138 research projects which were repair-related. Of these, only 60 per cent addressed problems identified from the case histories. There is plainly room for improvement here, in moving towards a more performance-related approach, and it is clear that site conditions must be taken into account more than they are at present; the difference between laboratory and site is currently too great. If this is not done, the very real potential benefits in building on the Principles in EN 1504 will not be realised. In his original paper Tilly [2.41] had made the following suggestions for improvement in moving forward:

(a) more attention to investigation and diagnosis of the original cause of the deterioration – its nature and extent.
(b) independent checking of the complete design process for the repair, to ensure compatibility and fitness of purpose in the prevailing environment.
(c) clear method statements and close supervision of the work in the field.
(d) greater use of testing, both for acceptance purposes, and in subsequent inspections.
(e) better training for site operatives.
(f) more education on the whole design process, to ensure better understanding in meeting clearly established performance criteria.

7.8 Timing of an intervention

There are various figures earlier in this book which illustrate different aspects of performance versus time. Figure 3.13 is fairly general with emphasis on load capacity, but carrying the suggestion in the text that this can be extended to cover different aspects of serviceability. Figure 4.8 takes this further, in defining different aspects of corrosion-related performance. Figure 3.3 introduces alternative strategies for strengthening bridges, based on practice in the UK.

There is a need now to consolidate this, in considering when to intervene, while making a choice of options from the five categories in Table 7.5. Figure 7.5 illustrates the possibilities. Figure 7.5(a) is general, showing a deterioration line in relation to the level of minimum acceptable performance.

Figure 7.5(b) relates to safety. While it is possible that this might simply involve maintaining the present condition, it is common to enhance strength to a level corresponding to the original undamaged state – or even higher. This might be done immediately, or, if confident, from the appraisal, of the gradient of the performance–time curve and reasonably content regarding the assessed minimum-performance line, at a later stage (shown as options (1) and (2) respectively in Figure 7.5(b)). Generally, this will involve a choice from categories 1 or 2 in Table 7.5.

If reasonably content regarding safety (always the prime concern), then performance is more general and the situation different. This is shown in Figure 7.5(c). Here, the main concern is to alter the gradient of the performance–time curve, to delay its arrival at the minimum-technical-performance line. This may be done for serviceability, functional or aesthetic reasons, or to reduce the risk of future structural deterioration. The performance requirements for the repairs are very different, as indeed are the methodologies, which will come from categories 3–5 in Table 7.5.

Figure 7.5 makes a very clear distinction between actions which involve altering the structure by directly affecting its capacity and those which do not. There are also practical reasons for doing this, since a strengthening repair has to be selected on the basis of a different set of criteria.

In terms of urgency of intervention, Figure 7.5 suggests that the choice of action will depend on how close the performance curve is to the minimum-performance line. If the safety level is considered to be unsatisfactory, then the owner may decide to act immediately or within a reasonable time (Figure 7.5(b)). If not, the choice may come from considerations in Figure 7.5(c), and, in general, these will more likely enter the picture, the higher the present condition point is on the performance–time curve.

This suggests a possible zonal approach to the timing of intervention and to its nature. The principles of this are shown in Figure 7.6, with an increasing need for structural repair in moving from zone 1 to zone 3.

To operate such a system, rigour is needed in the progressive appraisal process outlined in Chapters 5 and 6, to establish how much the various aspects of performance may have reduced compared with the as-built condition. The dividing lines between the zones are subjective to some extent and will depend on the owner's strategy for routine and preventative maintenance.

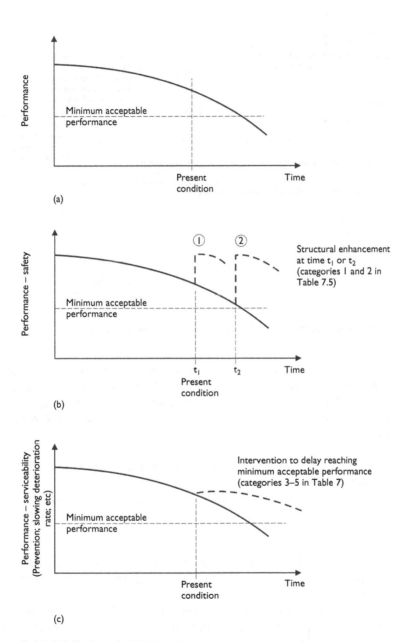

Figure 7.5 Performance–time curves

    (a) General
    (b) Structural enhancement/repair
    (c) Serviceability (prevention, etc.)

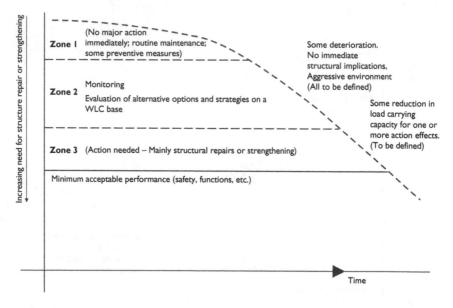

Figure 7.6 Suggested zonal approach to the timing of an intervention

Zone 1 represents an area of no immediate concern on performance terms. In zone 2, there is some deterioration, but, in general, no immediate structural implications. If the environment is particularly aggressive, monitoring may be introduced, but essentially this is an evaluation zone, in deciding:

–   either to introduce some form of preventative measure in the relatively short term.
–   or to allow further deterioration under controlled conditions, until residual load capacity becomes of concern, requiring a strengthening repair.

In zone 3, a structural repair is foreseen, the only issue being exactly when to carry it out before the minimum acceptable performance level is reached (Figure 7.5(b)).

Figure 7.6 is indicative only, and would have to be adapted for individual structural types and populations. In terms of timing of an intervention, it is included here to reinforce the point that repair categories 3–5 in Table 7.5 are likely to feature earlier in the life of a deteriorating structure than those in categories 1 and 2. As a sweeping generalisation, early proactive preventative measures will almost always be more economic, when evaluated in whole-life-costing (WLC) terms.

7.9   Selecting a repair option – general

7.9.1   Introduction

The selection of a repair option should be by a progressive process, rather like the assessment method recommended in earlier chapters. In assessment, the initial concern is with the whole structure (Figure 4.1), after examining previous records, before homing in, via testing and diagnosis, on the nature and extent of the deterioration – all leading to evaluations of condition and its effect on structural performance. The movement is from the general to the particular. Similarly in selecting a repair option, within an overall strategy, the first concern is with the general performance requirements for the repaired structure (Tables 7.1 and 7.2), before deciding what category of repair is necessary (Table 7.5) and then selecting an option from within the chosen category (Section 7.4). The progressive nature of doing that is shown in Figure 7.1, and the timing of the intervention is considered in Section 7.8, while making a clear distinction between structural enhancement and preventative measures.

This is all easier said than done. The lessons from performance of repairs in service (Section 7.7) tell us that the largely prescriptive approach in the past has led to a variable success rate, while giving some indication as to why that is so (Sections 7.7.2.6–7.7.2.8). The situation is further complicated by the stream of new options coming on into the market and frequent changes in the formulations of those that are already there, thus making comparison in evaluation difficult.

What is clear is that the selection of a repair option cannot be done in isolation, without consideration of how it is to be delivered – the roles of the various players in the supply chain, and how the whole process is managed, controlled and monitored. Section 7.7 tells us that this a sensitive process, and there has to be a shared understanding between all the parties of what is needed and what can be delivered, often driven by lowest first cost. The owner needs to buy into a whole process, and not just a particular repair system. There also has to be a shift away from a prescriptive approach towards one that is more performance-related and transparently shown to be so.

At the current stage of development of the repair industry, there is no universal panacea on how this can be achieved. While the industry will undoubtedly continue to evolve, there are clear signs of a more systematic and structured approach. The principal drivers for this are:

- the development of increasingly sophisticated asset management systems;
- increasing awareness of the need for a whole-life approach, both in terms of costs and performance;

- owners becoming more demanding in terms of their expectations both from the repair industry and the repairs themselves;
- the development of EN 1504 and its supporting Standards; and
- sustainability.

Details of progress in this area appear in a number of recent publications which go beyond the updating of traditional repair manuals. References [7.16–7.20] give a selection.

This brief outline suggests the flow chart shown in Figure 7.7, in selecting a repair option. Levels (1) and (2) are highly owner- and structure-specific, and coverage of all possible scenarios is beyond the scope of this chapter. Decisions here will be taken by the owner in consultation with his technical advisers.

Level (3) is structure-specific, and the decision on whether to go for option (b) or (c) in Figure 7.5 will depend not only on the owner's strategic plans and goals for the long-term future of the structure but also on the interpretation of the results from the assessment.

The major areas of selection are at level (4) in Figure 7.7, and the key issue is to identify selection tools which will help pinpoint the best-value option from the different categories in Table 7.5. It is important to remember that selection is relative rather than absolute, in making comparisons between options on some common basis. Whatever option emerges from this procedure then has to be subjected to the process implicit in level (5) in Figure 7.7; as emphasised earlier, this is to ensure a balance between the required performance and the delivery of that in practice.

7.9.2 Tools to aid selection

7.9.2.1 Introduction

At level (2) in Figure 7.7, a decision will have been taken to intervene, based on the inputs/issues at level (1), and the stage is set to move forward to level (3). Fundamentally, the choice is between a structural or non-structural option. Table 7.5 indicates what these are, together with some comments on possible application. Section 7.8 gives some guidance on timings, depending on the current position on the performance–time curve (Figure 7.6).

The selection process will be different for structural and non-structural options, largely because the performance requirements and criteria are different. The selection methodology will also be different, involving the use of different tools. Before looking at each of these basic categories of intervention, there is a need to be aware of what tools are available, and the

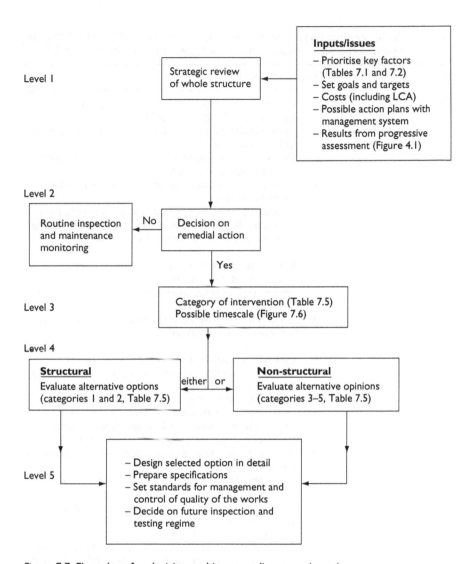

Figure 7.7 Flow chart for decision-making regarding a repair option

purpose of this section is to give brief details. More detailed information is available in references [7.12, 7.16 and 7.21].

In using these tools in practice, a perspective is necessary. In applying numerical methods, it is all too easy to get carried away by the apparent precision of the process. The validity and relevance of the input has to be borne in mind, as well as a host of practical issues. The CONREPNET report [7.16] is mindful of this, in listing out a series of barriers to achieving durable repairs; these are logged here in Table 7.7.

Table 7.7 Some potential barriers to achieving durable repairs – CONREPNET [7.16] (Reproduced with permission from reference 7.16. HIS, BRE Press, 2007)

Topic area	Potential Barrier or Issue
Design	• Owners' expectations and requirements not adequately defined or are unclear. • Incorrect design or specification of intervention from lack of knowledge. • Durability of intervention system/remediated structure not considered or known. • Insufficient attention given to the behaviour of the overall structure. • Lack of attention to design details or in specification of repairs, including the selection of materials, causing practical problems on site/lack of durability.
Durability	• Interventions are undertaken too late (structure excessively deteriorated). • There are unrealistic expectations of performance of interventions. • Various actors have significantly different expectations for intervention's durability.
Decision-making	• Lack of knowledge of relevant decision criteria. • Lack of knowledge of how to use available information. • That some owners have insufficient understanding of issues, risks and obligations. • That relatively high values of discount rate set by national governments produce unduly short-term perspective unsuited to long-life assets (e.g. bridges etc.) • WLC is rarely used, as most actors have an incomplete understanding of it. • Inappropriate choice of "key parameters" – lack of sensitivity to factors that really influence the production of durable interventions and remediated structures. • Quality and quantity of stored data inadequate for asset management purposes. • Financial restrictions limiting the ability of non-technical owners to buy-in appropriately skilled technical advice sufficiently early in the process.
Financial constraints	• Lack of budget for works, including a satisfactory level of site supervision. • Inadequate funding of maintenance. • Lack of appreciation by owners of value and contribution made by effective inspection, assessment and monitoring to successful durable intervention.

	• Consultancy input appearing to be expensive compared with "tangible results". Owners not being convinced of the need to make this investment.
	• An unduly rigid adherence to acceptance of cheapest tenders rather than those offering the best longer-term value for money.
	• Lack of accepted "financial" practice for establishing "best value" options.
Poor workmanship/ working practice	• That the people responsible for the technical assessment of the problem may not commission the intervention. Problems of interpretation and coherence in transfer.
	• Poor workmanship due to lack of proper supervision of site work.
	• That the people commissioning interventions may be administrators or general engineers not experienced in the work and act without suitable technical advice.
Education and training	• Lack of comprehensive guidance in a single document or suite of straightforward guides, which is neither unduly technical nor superficial.
	• Lack of training at all stages of the repair process. There is a need to understand the materials, their properties and limitations in their use and performance.
Statutory requirements, contract forms, guarantees, warranties and insurance issues	• Identification of relevant statutory requirements and interpretation on how these are met.
	• Availability of suitable contract forms for various potential collaborative working relationships or the competitive selection of technical solution/service provider.
	• Absence of an appropriate and commonly accepted guarantee/insurance period.
	• Availability of suitable guarantees/insurance facilities for proposed methods of project working and potential difficulties arising in some European countries.
Material issues	• The substitution of alternative and cheaper material which notionally comply with the specification, but do not provide the same durability or performance.
	• The large number of materials on the market making comparison very difficult.

Table 7.7 (Continued)

Topic area	Potential Barrier or Issue
	• A reluctance to specify and use new materials because of the risk of poor performance and liability incurred as a result. Risk and benefits sharing needed. • Standards which allow material to be tested in different ways (alternative test methods) so producing incompatible results. Need for harmonisation process producing consistency and compatibility in test outputs. • Lack of performance criteria and testing regimes creating the potential for disagreement between owner/consultant/contractor/other as to whether the performance of a material complies with a specification etc. • Extremes of weather during the remedial work causing problems with durability.

In considering the different selection tools, it is also worth remembering that the ideal situation would be a proven track record under field conditions corresponding to those pertaining to the structure being evaluated. As with conventional structural design, this utopian situation will occur only very rarely, but the selection process should be shaped to get as close as possible. This requires clarity in defining the starting conditions in terms of:

• being clear on the primary deterioration mechanisms;
• establishing the nature, scale and location of the deterioration;
• defining the local micro-climate and environmental conditions.

7.9.2.2   *Essential elements of selection tools*

An examination of Table 7.5 reveals that the categories of repair options are widely varying, thus making the derivation of a common base for comparison purposes somewhat difficult. Most work in this field has been done on categories 3–5, while including a general classification of patch repairs, without always distinguishing between those that are structural and those that are not. In identifying essential elements for selection tools, therefore, it is necessary to concentrate on these categories.

In moving away from a purely prescriptive approach to one that is more performance-based, the essential elements of any tool are:

 (i) identification of indicators (performance requirements);
 (ii) definition of threshold values for these;
(iii) consideration of execution/delivery issues;
(iv) the use of acceptance testing and verification;
 (v) developing a system for post-intervention monitoring.

The core of the methodology lies in elements (i) and (ii), in assembling all essential performance requirements and in setting limiting values (thresholds) for these. In doing this, some tools under development can be quite complex, with skill required in setting weighting factors for each indicator and targets for the outputs from the multi-attribute evaluation process. Phases (iii)–(v) then involve the introduction of modification factors to the basic outputs, to take account of execution and quality issues, as well as to take the micro-climate into consideration.

Before homing in on this with respect to repair options and their performance requirements as given in Section 7.4, it is worth recognising that the methodology is simply a systematic approach to evaluation and decision-making, which might also be used with advantage when considering the issues at level (1) in Figure 7.7. There, the owner has to review the long list of performance requirements given in Table 7.1, while prioritising these and setting targets. In then considering the nature of possible intervention, similar thinking could be applied to the more structure-related requirements in Table 7.2.

To illustrate this, Table 7.8 contains a selection of performance requirements taken from Table 7.1. For each of these, some performance indicators are suggested, without in any way being definitive or comprehensive. The owner would make his own choice, in terms of what is especially important to him. He would then apply a weighting factor to each indicator, to some pre-determined scale, to arrive at a score, which could indicate both the urgency of the repair and its nature. The actual methodology and scoring system would depend on the favoured selection method (see Section 7.9.2.3). However done, the end result should be a repair index (RI) given by

$$RI = \sum_{1}^{n} (PI) \times (WF)$$

Where PI are the relevant performance indicators
WF are the corresponding weighting factors and
n is the number of PIs.

The method is empirical, requiring a great deal of judgement. However, it is a simple numerical approach, which attempts to bring some order to a complex multi-attribute evaluation process. In cases where it has been applied in the past, most weight has been put on economy and durability,

Table 7.8 Illustration of a simple ranking approach to determine the urgency and nature of intervention

A selection of performance requirements from Table 7.1	Some possible performance indicators (PI)	Ranking/weighting factors (WF)
1 Safety	Consequences of failure Type of failure Robustness (structural sensitivity) Reliability of repair	For each PI, one possibility is: Very high 4 High 3 Medium 2 Low 1
2 Serviceability, function, aesthetics	Stiffness and deformation Continuity of operations Appearance Inspectability	The object is to determine a "score", which indicates the urgency of intervention, while giving some indication of its nature.
3 Durability	Local micro-climate Nature and number of aggressive actions Extent and nature of deterioration Projected life of repair system	A high score denotes greater urgency.
4 Economy	First costs Whole life costs User costs	
5 Health and Safety	Public safety Safety during reconstruction	
6 Sustainability and environmental issues	Use of sustainable materials Control of emissions, noise, etc. Energy consumption	

but the trend must be towards greater weight on safety, serviceability and sustainability.

The first four performance requirements in Table 7.8 are most closely related to deciding the urgency and nature of the repair, while giving some idea of the expectations of the remedial action. As such, this stage is no more than a lead-in phase, to give focus to the subsequent, more detailed phase of repair selection which is considered next, using patch repairs as an example.

In this chapter, patch repairs are considered in Section 7.4.3, and classified as structural repairs in Table 7.5. In general, sufficient material is removed, which will temporarily reduce the strength and stiffness of the element, and this aspect has to be carefully considered in planning the work. Once again, the importance of pre-preparation is stressed. The fundamental performance requirement of the repair is to ensure that the repaired element acts as a composite entity, which requires compatibility between the original concrete and the repair material, as well as effective bond with both the concrete and the pre-prepared reinforcement.

Basic performance indicators are given in Section 7.4.3, and, in making a selection from the repair options available in the market, it is necessary to establish the relative important of these. The CONREPNET report [7.16] illustrates one way of doing that, as illustrated in Table 7.9. Three performance requirements are given, and, for each, performance indicators have been selected, together with a judgement of their relative importance. (I_m, where the total weighting should add up to 1.0.) Column 5 sets some provisional performance criteria, with the corresponding scores given in column 4. An Intervention Performance Index (IPI) is then obtained by multiplying the weighting (column 3) by a best judgement of the score for column 4. Thus, the score to date in column 6 is given by

$$IPI = \sum PI.I_m$$

and totalled in the bottom right-hand corner of the table. This would be compared with a pre-determined target value, which, for the system shown in Table 7.9, should be taken as 3.0. As illustrated, the method gives a measure of the need for repair ($\Sigma\ PI.I_m \geq 3.0$), but the approach can be used to compare different patching systems.

The approach can also be used to compare different protective and remedial measures. In this broader context, it is also necessary to introduce a modifying factor which takes some account of the effect of the aggressivity of the local environment and its possible effect on the durability of alternative repair methods. The CONREPNET suggestion for doing this is to use the environmental classification system now given (with minor variations) in most design Codes and Standards – shown typically in Table 7.10 for Eurocode 2 [7.22]. The proposal is that an environmental classification factor (EA) be developed, which attributes a score related to the expected performance of each repair method in each of the class designations in Table 7.10. The approach is to adopt the same score ranges as those in Table 7.9, i.e. 1–4, with 4 being a poor performance and 1 being good. The CONREPNET suggestions for this are given in Table 7.11, where it may be seen that cathodic protection and coatings are generally less sensitive to the environment.

Table 7.9 Illustration of performance indicator approach for patch repairs – CONREP-NET [7.16] (Reproduced with permission from reference [7.16]. HIS, BRE Press, 2007)

PR	PI	Im_{PI}	Score	Performance Criteria (Provisional)	Score at Date...
Performance Requirements	Performance Indicator	Relative Importance (Weight $\sum Im_{PI} = 1$)			
Bond	Debonding : Crack width between repair and substrate (visual)	0.05	4 3 2 1	>0.4 mm 0.1–0.4 mm 0.05–0.1 mm <0.05 mm	$PI_1.Im_{PI1}$
	Bond to substrate	0.10	4 3 2 1	<0.3 MPa 0.3–0.7 MPa 0.7–1.1 MPa >1.1 MPa	$PI_2.Im_{PI2}$
Permeability	Cracking in the patch material	0.10	4 3 2 1	>0.4 mm 0.1–0.4 mm 0.05–0.1 mm <0.05 mm	$PI_3.Im_{PI3}$
	Carbonation front rate factor	0.10	4 3 2 1	>6 mm/year$^{0.5}$ 3–6 mm/year$^{0.5}$ 1–3 mm/year$^{0.5}$ <1 mm/year$^{0.5}$	$PI_4.Im_{PI4}$
	Chloride ion diffusion coefficient	0.10	4 3 2 1	>5 × 10^{-12} m^2/s 2–5 × 10^{-12} m^2/s 1–2 × 10^{-12} m^2/s <1 × 10^{-12} m^2/s	$PI_5.Im_{PI5}$
	Water absorption or sorptivity	0.07	4 3 2 1	>0.2 mm/mm^2 0.15–0.2 mm/mm^2 0.1–0.15 mm/mm^2 <0.1 mm/mm^2	$PI^6.Im_{PI6}$
Durability	Reinforcement potential	0.15	4 3 3 2 1	>−350 mV (SCE) −250/−350 mV (SCE) −250/−350 mV (SCE) −100/−250 mV <−100 mV (SCE)	$PI_7.Im_{PI7}$
	Reinforcement corrosion rate	0.15	4 3 2 1	>10 μm/year 5–10 μm/year 1–5 μm/year <1 μm/year	$PI_8.Im_{PI8}$

Concrete cover resistivity	0.10	4 3 2 1	$<50\,\Omega.m$ $50–100\,\Omega.m$ $100–500\,\Omega.m$ $>500\,\Omega.m$	$Pl_9.Im_{Pl9}$
Concrete cover mechanical strength	0.08	4 3 2 1	$<10\,MPa$ $10–20\,MPa$ $20\,MPa$ – Parent concrete $>$Parent concrete	$Pl_{10}.Im_{Pl0}$ $\sum Pl.Im_{Pl}$

Table 7.10 Environmental classification from Eurocode 2 [7.22], concerned with the corrosion of reinforcement (Permission to reproduce extracts from the British Standards is granted by BSI)

Class designation	Description of environment	Informative examples where exposure may occur
1. No risk of corrosion or attack		
X0	For concrete without reinforcement or embedded metal: all exposures except where there is freeze/thaw, abrasion or chemical attack. For concrete with reinforcement or embedded metal: very dry.	Concrete inside buildings with very low air humidity.

2. Corrosion induced by carbonation

Where concrete containing reinforcement or other embedded metal is exposed to air moisture, the exposure shall be classified as follows:

Note: The moisture condition relates to that in the concrete cover to reinforcement or other embedded metal but, in many cases conditions in the concrete cover can be taken as reflecting that in the surrounding environment. In these cases classification of the surrounding environment may be adequate. This may not be the case if there is a barrier between the concrete and its environment.

XC1	Dry or permanently wet	Concrete inside buildings with low air humidity Concrete permanently submerged in water

Table 7.10 (Continued)

Class designation	Description of environment	Informative examples where exposure may occur
XC2	Wet, rarely dry	Concrete surfaces subject to long-term water contact Many foundations
XC3	Moderate humidity	Concrete inside buildings with moderate or high air humidity External concrete sheltered from rain
XC4	Cyclic wet and dry	Concrete surfaces subject to water contact, not within exposure class XC2

3. Corrosion induced by chlorides other than from sea water

XD1	Moderate humidity	Concrete surfaces exposed to airborne chlorides
XD2	Wet, rarely dry	Swimming pools Concrete exposed to industrial waters containing chlorides
XD3	Cyclic wet and dry	Parts of bridges exposed to spray containing chlorides Pavements Car parks

4. Corrosion induced by chlorides from sea water

XS1	Exposed to airborne salt but not in direct contact with sea water	Structures near to or on the coast
XS2	Permanently submerged	Parts of marine structures
XS3	Tidal, splash and spray zones	Parts of marine structures

Table 7.11 CONREPNET [7.16] proposals for EA ratings for various repair methods, based on the environmental classes in Table 7.10 (Reproduced with permission from reference [7.16]. HIS, BRE Press, 2007)

Repair method	EN 206 Environmental class										
	X0	XC1	XC2	XC3	XC4	XD1	XD2	XD3	XS1	XS2	XS3
Patching	1	1	1	2	3	2	3	4	2	3	4
Coating	1	1	1	1	2	1	2	3	1	2	3
Crack injection	1	1	1	2	3	2	3	4	2	3	4
Sprayed concrete	1	1	1	2	3	2	3	4	2	3	4
Cathodic protection	1	1	1	1	2	1	2	3	1	2	3

It is then proposed that a new Index be obtained by adding the EA result to the basic IPI value. This requires a judgement on the relative importance of the two, and the CONREPNET proposal is that 20 per cent be assigned to the EA value. Thus:

$$\text{Intervention Life Index (ILI)} = 0.8\,\text{IPI} + 0.2\,\text{EA}$$

Finally, and reverting to patch repairs in particular, where there are many different systems on the market with varying degrees of difficulty in execution and/or necessary levels of quality control, some allowance needs to be made for these practical issues. It is difficult to see how this may be done in any meaningful way using a numerical approach to qualify selection. This can only be done by recognising the need in the totality of the design of the repair system, and giving a lot of attention to:

- the specification
- pre-preparation of the substrate
- clear method statements
- levels of supervision
- training of both engineers and operatives

The methodology described here is far from rigorous, but an attempt to bring some order to selection and to be more objective and performance-related. Particular systems, based on these principles, are itemised and referenced in Section 7.9.2.3.

7.9.2.3   Current methods under development

The REHABCON Manual [7.21] contains a series of appendices which review the repair categories in Table 7.5. The Manual itself presents a comparatively simple method for making a selection from these; this is also presented in reference [7.23]. However, most detail on the selection tools briefly listed in this section, is obtainable from references [7.16] and [7.19].

More generally, we enter the world of acronyms, as follows:

(a)   MADA – Multi-Attribute Decision Aid
This is a device for decision making, where there is more than one alternative to choose from and more than one criterion for each alternative. In the LIFECON project [7.19], three variations were applied, differing by the techniques used to determine weighted attribute values; these ranged from a simple ranking approach to a more complex definition of threshold values.

(b) QFD – Quality Function Deployment

QFD has its origins in the 1950s when used by industry for product development and is essentially a matrix method which links requirements to process properties via weighting factors. It is thus a quantitative approach as outlined in Section 7.9.2.2, and when linked to another (qualitative) tool RAMS (Reliability, Availability, Maintainability, Safety) can address the planning of maintenance interventions. An attractive features is an integrated sensitivity analysis.

(c) LMS – Life-cycle Management System for concrete structures As the name implies, this is not an individual tool, but a comprehensive approach to the asset management of concrete structures, developed during the LIFECON project [7.19]. It involves a number of inter-related modules. Stemming from a basic concern for life-cycle cost analysis, use is made of both MADA and QFD, together with a range of risk-analysis techniques. The scope is wide-ranging, embracing degradation mechanisms, condition of the structure, maintenance and repair methods – all focused on life-cycle action plans and decision-making.

No attempt is made here to describe these tools in detail; the reader is referred to references [7.16 and 7.19], where the methodology is described together with worked examples. They are referenced here simply to make the point that the essential elements of any tool, outlined in Section 7.9.2.2, can be, and have been, developed into quite detailed numerical methods for selection purposes. In using them, it is essential to retain a sense of perspective, as indicated in Section 7.9.2.1.

7.10 The role of EN 1504 [7.13] in selection

Table 7.3 indicates that the ten parts of EN 1504 have all been published, with two having undergone revision. Collectively, these cover the 11 Principles for Repair given in EN 1504-9 and briefly summarised in Table 7.4; of these, the first six are general and the last five relate to different aspects of corrosion resistance. The ten parts do not correspond to the 11 principles, as may be seen from the expansion of Table 7.3 given in Table 7.12. For example, aspects of surface coatings are covered in EN 1504-2 and EN 1504-7, while other Parts deal with particular stages of the overall process, e.g. Part 9 on design and Part 10 on execution.

Table 7.12 has been downloaded from the Concrete Repair Association website [7.30], which also contains a series of fact sheets which help relate the different principles to the different options listed in Table 7.5.

The scope of EN 1504 is comprehensive, and provides a more performance-related approach to repair. The key features within that scope are:

Table 7.12 Outline of the contents of Parts 1–10 of EN 1504 (reference [7.30])

CEN Standard	Title and brief description	Anticipated Publication date
BS EN 1504-1:2005	Definitions Recently revised, following completion of other parts of BS EN 1504, this Part gives common definitions used in the EN 1504 series of standards.	Published 2005
1504-2	Surface protection systems: specifies requirements for the identification, performance (including durability aspects) and safety of products and systems to be used for surface protection of concrete, to increase the durability of concrete and reinforced concrete structures, as well as for new concrete and for maintenance and repair work.	Published October 2004
1504-3	Structural and non-structural repair: specifies requirements for the identification, performance (including durability) and safety of products and systems to be used for the structural and non-structural repair of concrete structures.	Published February 2006
1504-4	Structural bonding: specifies requirements for the identification, performance (including durability) and safety of structural bonding products and systems to be used for the structural bonding of strengthening materials to an existing concrete structure, including: – bonding of external plates of steel or other suitable materials (e.g. fibre reinforced composites) to the surface of a concrete structure for strengthening purposes, including the laminating of plates in such applications. – bonding of hardened concrete to hardened concrete, typically associated with the use of precast units in repair and strengthening. – casting of fresh concrete to hardened concrete using an adhesive bonded joint where it forms a part of the structure and is required to act compositely.	Published November 2004
1504-5	Concrete injection: specifies requirements and conformity criteria for identification, performance (including durability aspects) and safety of injection products for – force transmitting of cracks, voids and interstices in concrete (FTFC), – ductile filling of cracks, voids and interstices in concrete (DFC), – swelling fitted filling of cracks, voids and interstices in concrete (SFFC)	Published December 2004

Table 7.12 (Continued)

CEN Standard	Title and brief description	Anticipated Publication date
1504-6	Grouting to anchor reinforcement or to fill external voids: specifies requirements for the identification, performance (including durability) and safety of products to be used for the anchoring of reinforcing steel (rebar) as part of the repair and protection of concrete structures.	Published September 2006
1504-7	Reinforcement corrosion protection: specifies requirements for the identification and the performance (including durability aspects) of products and systems for active and barrier coatings for protection of existing steel reinforcement in concrete structures under repair.	Published September 2006
1504-8	Quality control and evaluation of conformity: specifies procedures for quality control and evaluation of conformity, including marking and labelling of products and systems for the protection and repair of concrete according to Parts 2–7 of EN 1504.	Published November 2004
DD ENV 1504-9:1997	General principles for the use of products and systems: specifies basic considerations for specification and protection and repair of plain and reinforced concrete structures with products and systems which are specified in the EN 1504 series or standards or any other relevant EN or European Technical Approval (ETA).	Published July 1997 under conversion to EN 1504-9
BS EN 1504-10	Site application of products and systems and quality control of the works: gives requirements for substrate condition before and during application including structure stability, storage, the preparation and application of products and systems for the protection and repair of concrete structures including quality control, maintenance, health and safety, and the environment.	Published March 2004
Standards produced by other committees, relevant to protection and repair		
BS EN 12696:2000	Cathodic protection of steel in concrete.	Published March 2000
Pr EN YYY	Sprayed concrete for repair and upgrading of structures.	Not known
prEN 14038-1	Electrochemical re-alkalisation and chloride extraction treatments for reinforced concrete – Part 1 : Re-alkalisation.	Not known – may not be published
prEN 14038-2	Electrochemical re-alkalisation and chloride extraction treatments for reinforced concrete – Part 2 : Chloride extraction.	Not known – may not be published
EN 206-1:2000	Concrete – Part 1 : Specification, performance, production and conformity.	Published

(a) the definition of performance requirements
(b) the provision of a design basis
(c) coverage of all aspects of execution
(d) definition of procedures which permit evaluation of conformity for products and systems, underpinned by a raft of supporting Standards on test methods. A selection of these is given in Appendix 7.1 for CEN Standards and in Appendix 7.2 for ISO standards.

It is also clear that the repair industry is responding positively to this initiative, and EN 1504 will play an increasing role in the selection process in the future, particularly for all protection options and general patch repairs. This is encouraging, since the approach is more physically related to the materials and techniques for repair, and to execution on site.

7.11 Selecting a repair option in practice

Section 7.9 gives a brief general review of current approaches to selection, together with references where further details may be found. This section looks at these in the context of Figure 7.7, while focussing on levels (3), (4) and (5). In practice, the use of selection tools is only part of a wider process, which also embraces technical and functional issues.

7.11.1 Level (3) in Figure 7.7 – Deciding on a structural or non-structural intervention

At this level, the decision will have already been taken that some action is necessary, and this has to be extended to decide on the nature of that. This is a fundamental decision, rather than a pure selection process. Strategic issues at level (1) will dictate this decision, while being mindful of the performance requirements in Table 7.1 and of where the current position is located on the performance–time curve (Figures 7.5 and 7.6). If in zone 3 in Figure 7.6, then a structural repair will be the front runner. If in zone 2, then either a structural enhancement or a preventative approach might be chosen. With the focus now firmly on the structure itself, the factors which influence this decision include the following:

(a) Owner attitude, in terms of how early to move when deterioration is indicated, i.e. his perception of minimum acceptable technical performance within his overall management and maintenance strategy.
(b) Costs. This relates both to the capital cost of the repair, and to whole-life costs, which in turn are related to the estimated life and efficiency of the available options. In general, preventative measures are cheaper than structural options in terms of first costs, but their effective life may be shorter.

(c) The nature and depth of the assessment process, and the confidence in the results obtained. This covers not only the current structural position, but also a clear identification of the dominant deterioration mechanism and its predicted future rate, i.e. not just the present position on the performance–time curve, but also the future gradient of that curve.

(d) The extent and nature of the damage, and also its location, i.e. is it close to critical sections?

(e) Structural sensitivity, i.e. its general robustness and the nature of the most critical local effect. Potential deficiencies in shear or punching shear, or the likelihood of a brittle failure in general, prompt earlier structural intervention.

(f) An assessment of the real imposed loads on the structure and confidence in any more rigorous analysis used to determine the effects of these. Current real loads may be less then those assumed in design, but future loadings may be higher within the projected life of the structure.

(g) The nature of the structure and ease of access to the deteriorated elements. Some options are ease of access to the deteriorated elements. Some options are easier for certain types of structure, e.g. protective overlays on bridge decks, whose nature may also permit some forms of external strengthening.

There are no general applicable rules for taking a decision at level (3), but it is of great importance to clearly decide on whether the repair is to be structural or non-structural, since the selection processes for each are different at level (4). Of the seven factors listed above, (a) and (b) will normally provide the starting point, in relation to the zonal approach in Figure 7.6. Factors (c)–(g) are primarily technical, in the follow-through to a decision on the category of repair and the timing of the intervention.

7.11.2   Level (4) in Figure 7.7 – Structural repairs

The options are those listed in categories 1 and 2 in Table 7.5. When it comes to selection, the external options in category 1 need to be re-classified as being either genuine strengthening solutions (options (a), (b) and (c)) or barriers (options (d), (e) and (f)).

7.11.2.1   External strengthening solutions

The three options are:

(1) replacing complete elements
(2) strengthening via plate bonding or some form of confinement
(3) external post-tensioning

In all three cases, technical selection will be based on calculations, first in assessing the residual capacity of the unrepaired element and second in justifying how that will be enhanced by the different options. Final selection will then depend on costs and on aptness for the type of structure involved. The numerical approaches in Section 7.9 will have only a minor role, if any, in selection.

Option (1) is straightforward. A new element is designed in accordance with current Codes and Standards, requiring additional consideration of how to fit it into the remaining structure.

Option (2) is more complex, since there are more alternatives available. Historically, the use of externally bonded steel plates began about 40 years ago, and there are examples of the technique in many countries, particularly in Japan with several thousand applications, mainly for bridges. Design guidance does exist in the UK [7.24], based on reviews of extensive testing, e.g. references [7.2, 7.3, 7.6] and [7.25]. A particular concern is the long-term performance of the adhesives used for bonding, and the evolution of adhesives is described in references [7.25 and 7.26].

Plate bonding has mainly been used to strengthen beams and slabs in bending and shear, with careful detailing required for the ends of the plates. The choice of adhesive is also important, as are joint thickness and the method used to offer up the plate to the concrete surface. In selecting this option, assurance would be required from test data on:

(i) the added strength, assuming full composite action; and
(ii) the suitability of the adhesive, both in the short and long term.

In recent years, external strengthening has moved on, with the use of fibre reinforced composites becoming more common. In sheet form, these can be used as direct replacements for steel plates. In sheet or strap form, they are also used in a wrap-around confinement role, particularly for strengthening columns. Reference [7.6] covers this technique in design terms, and it is interesting that it is complemented by reference [7.4], covering acceptance inspection and monitoring – a particular example of what should be a universal trend for all repair systems.

There are 174 references to the Concrete Society Technical Report 55 [7.6], nearly all published in the past decade, which give a clear indication of the depth of development work that has been done on the world stage. The target design life is 30 years. The detailed design approach is the traditional one of partial factors, applied to bending, shear and compression, with limits set on elongations in some applications.

The fibres used derive from carbon, aramids or glass. The adhesives are generally epoxies, but can be polyesters, vinyl esters or polyurethane. Fibre composites have a straight-line stress–strain curve with no yield point;

Table 7.13 Indicative dry fibre properties of fibre composites for strengthening work (Reprinted by permission of The Concrete Society. All rights reserved [7.6])

Fibre	Tensile strength (N/mm²)	Modulus of elasticity (kN/mm²)	Elongation (%)	Specific density
Carbon: high strength[a]	4300–4900	230–240	1.9–2.1	1.8
Carbon: high modulus[a]	2740–5490	294–329	0.7–1.9	1.78–1.81
Carbon: ultra high modulus[b]	2600–4020	540–640	0.4–0.8	1.91–2.12
Aramid: high strength and high modulus[c]	3200–3600	124–130	2.4	1.44
Glass	2400–3500	70–85	3.5–4.7	2.6

[a] based on polyacronitrile precursor
[b] based on pitch precursor
[c] aramids with the same strength but a lower modulus are available but are not used in structural strengthening applications.

indicative properties are given in Table 7.13 [7.6]. Elastic methods of analysis are therefore used, without re-distribution. Normal section design applies, but with checks to avoid peeling failures and debonding from the concrete. These materials have advantages over steel plates in terms of lightness and high strength, as well as being easier to install without need of temporary supports.

The companion Concrete Society guidance document on acceptance, inspection and monitoring of strengthening with fibre composites [7.4] also covers all aspects of testing, with a strong emphasis on the EN 1504 methodology, including the keeping of records both during installation and subsequent inspection and monitoring. The importance of this cannot be over-emphasised; it is a total design, installation and maintenance package, essential to the achievement of the desired design life. Table 7.14 gives a flavour of the approach, on the particular aspects of checks during installation.

Finally, there is the third strengthening option in category 1 in Table 7.5, the use of external post-tensioning. This is a technique that has been used for decades to provide additional strength to structural elements in all sorts of materials. The principal use has been for additional strength in bending and shear, using a variety of cable profiles as shown in simple form in Figure 7.8.

The basis of design is traditional, e.g. via Codes of Practice [6.2, 6.17], with more detailed guidance available for particular applications [7.27].

Table 7.14 Summary of main points to look for, during the installation of fibre composites (Reprinted by permission of The Concrete Society. All rights reserved [7.4])

Materials
- Were all materials (plates, fabrics, adhesives, resins, etc.) in accordance with the specification?
- Were certificates of conformity supplied with the materials?
- Were all materials handled and stored according to manufacturers' guidelines?

Surface preparation
- Was the concrete surface prepared in accordance with the specification?
- Was the surface regularity in accordance with the specification?
- Where applicable, were the corners of the concrete elements rounded?
- Was the surface of FRP plates prepared in accordance with the specification?

Adhesive
- Was the correct ratio of components used?
- Was the colour uniform (within a batch and between batches), indicating consistent mixing?
- Was application completed before the end of the open time?
- Was the adhesive thickness controlled? If spacers were used, were they at the specified locations?
- Was the structure subject to significant vibration while the adhesive was curing?
- Where adhesive thickness varied (e.g. where plates cross) did the plates spring away from the surface?

Plates
- Were the plates correctly oriented?
- Were plates moved after first coming into contact with the concrete?
- For carbon-fibre plates, was there any contact with metallic parts?

Fabrics
- Was the right amount of resin used?
- Was the fabric correctly oriented?
- Were there any wrinkles or irregularities in the fabric after compaction?
- Was the fabric moved after coming into contact with the concrete or the previous layer?
- Was the quality of each layer checked before the subsequent layer was installed?
- For carbon fibre fabrics, was there any contact with metallic parts?

Curing
- Was the FRP system correctly cured?

Testing
- Were test specimens prepared in accordance with the requirements?

Inspection
- Was the completed strengthening inspected for voids by tapping or other means?

Records
- Were the appropriate records maintained?

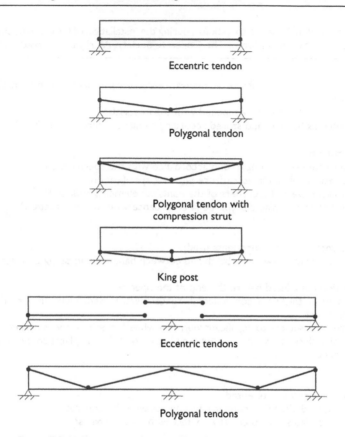

Eccentric tendon

Polygonal tendon

Polygonal tendon with
compression strut

King post

Eccentric tendons

Polygonal tendons

Figure 7.8 Indicative tendon profiles for use in strengthening by external post-
tensioning

Particular care is necessary in the design and detailing of anchorages and
deviators, to ensure effective transfer of the prestress to the parent structure, while avoiding excessive local stresses in the concrete. Tendons will
normally require lateral restraint. The act of introducing additional prestress into an existing structure induces strains and deformations, both local
and general. The whole question of movement requires attention in design,
particularly in the presence of restraints. As for all major strengthening
options, standards of execution and workmanship are of great importance;
for post-tensioning in the UK, suitable standards and practices are available
[7.28].

The three major strengthening options briefly described above are all
engineering-orientated, driven by structural design. Each has its own characteristics and all are fairly well-established both in terms of design and
execution. In making a choice, the use of the numerical tools listed in Section
7.9 are of little help. Technically, selection will depend on:

- the extent and nature of the deterioration;
- whether the objective is to restore to the original structural capacity or to seek further enhancement;
- the nature of the structure, and its suitability for the alternative options;
- confidence in the reliability of the option, as determined by analysis and calculations.

Above all, selection will be governed by the relative costs of each, and normal practice would involve cost comparisons, in consultation with contractors experienced in the different execution procedures.

7.11.2.2   External barrier solutions [(d), (e) and (f) in category 1: Table 7.5]

These three options are essentially preventive in nature but were included in category 1 in Table 7.5, because they all involve additional positive external barriers, beyond that provided by surface treatments or coatings (category 3).

It is quite clear, both from the literature and practice, that the rate of deterioration, due to a dominant mechanism such as corrosion, can be slowed significantly by controlling the moisture state at concrete surfaces, by a combination of barriers and good drainage. The waterproofing of bridge decks has been standard practice in the UK for many years, augmented in recent times by the concept of bridge enclosures to control the local mirco-climate.

There are many ways of doing this, and it is not proposed to go into detail. Most of the experience stems from the world of bridges, and the relative merits of acceptable alternatives may be assessed by examining the options contained in the Design Manual for Roads and Bridges (DMRB) [7.29] It is likely, however, that options in this category will be more expensive that those in category 3 (Table 7.5), at least in first cost terms.

7.11.2.3   Internal Strengthening Solutions (Category 2, Table 7.5)

Category 2 is defined as a structural repair. The literature refers to a general category of 'patch repairs', without always making a distinction between filling in relatively small holes with the prime objective of reinstating protection and making repairs which enhance structural capacity. For the vast majority of situations, the materials and techniques used may largely be the same with similar performance requirements – but the objective of the repair is clearly different and will be influential in making a choice between the numerous proprietary systems on the market. In addition, some types of repair are uniquely structural, e.g. locating and grouting dowel bars into pre-drilled holes to enhance shear strength.

In terms of available guidance, Concrete Society Report 38 [7.8] relates specifically to patch repairs used to reinstate protection, with a stated maximum area of 0.5 m² and a depth of less than 100 mm; the bulk of the report is devoted to a model specification. More generally, repair manuals (e.g. reference [7.18]) devote a great deal of space to available materials and systems and to the practicalities of execution. Up-to-date information on these aspects is also available on websites [7.30].

For a structural patch repair, priority will be given to options which generate a high bond strength to ensure composite action between the old and new concrete, and to provide sufficient bond and anchorage for the reinforcement. This need may be less important where the patch is intended simply to reinstate protection, although adhesion between the two concretes still matters a great deal. For both situations, additional protection to the patches is often provided by a surface coating.

Cementitious or polymer-modified materials are used for patch repairs in practice, and an important characteristic is that they should be low-shrinkage. In feedback from practice, Figure 7.2 clearly shows the dominance of corrosion, cracking and debonding as failure modes, and the design of the repair should minimise the risk of these occurring.

Section 7.4.3 lists some of the desirable characteristics for repair materials. Table 7.9 is a particular example of how to rank these as part of the selection process, but it should be noted that the indicators – and their relative importance – will vary, depending on whether the primary objective is structural or protective. As indicated in Table 7.7, the large number of different materials and systems on the market makes comparisons between them rather difficult, but the general approach outlined in Section 7.9 is a useful starting point in selection, but with the trend towards methods based on EN 1504 (Section 7.10).

BS EN 1504-3: 2004 [7.13] is useful in setting out the performance requirements for repair mortars and concretes, including making a distinction between those to be used for structural and non-structural repairs. It also makes recommendations for different application methods (by hand, spraying, recasting etc.) and sets out quality control and acceptance procedures.

A clear message from the feedback provided in Section 7.7 is that the success rate for patch repairs is dependent on workmanship and the general standards of execution on site. This stresses the importance of method statements and levels of supervision, linked to the greater use of testing within clearly established acceptance procedures – in fact, all the factors listed in Section 7.7.3. However, these improvements will still be negated, if the standard of pre-preparation is poor.

Pre-preparation begins at the assessment stage, in identifying not only the deterioration mechanism but also the full extent of its effects and whether or not it continues to be active. This is particularly true of corrosion, in understanding the type of corrosion cell involved and in ensuring that the corrosion activity is not simply moved away from the planned repair area. This suggests a conservative approach in removing both deteriorated concrete and rust products, i.e. taking out some sound concrete as well.

This then suggests the following essentials, in preparing the substrate:

1   Generous removal of contaminated materials and shaping of the area to be patched, e.g. avoid feathered edges
2   Thorough cleaning of the substrate (dust, dirt, grease, oil, etc.)
3   Avoiding micro-cracking
4   Providing reasonable roughness to the prepared surface of the concrete
5   Pre-watering of the surface, but this should be superficially dry at the time of casting.

Various techniques, including water-jetting, sand- and shot-blasting and the use of jack hammers, have been used to remove contaminated concrete. Some research has been done in Sweden [7.31] on the relative merits of these in terms of minimising damage to the residual substrate and generating potential adhesion strengths of up to 1.5–2.0 Mpa. This showed that water-jetting and sand-blasting created no micro-cracking, with water-jetting being preferred for its selective removal capabilities and causing least damage to the reinforcement.

With respect to the reinforcement, existing corrosion has to be eliminated or neutralised. Protection against future corrosion is also important, involving the use of primers which may be simple barriers, cement-based products or inhibitor coatings.

Full coverage of all the options on the market is beyond the scope of this sub-section; these are many and varied. What is important is that the repair is properly designed, leading to full specifications and method statements, which in turn are supported by good quality-control and agreed acceptance procedures. Sections 7.9 and 7.10 indicate that selection tools are available for this category of repair, whilst, in this sub-section, emphasis has been put on the importance of site practice and execution.

7.11.3   Level (4) in Figure 7.7– Non-structural repairs

The options are categories 3 and 4 in Table 7.5:

(a)   Surface coatings and impregnations
(b)   Filling cracks and voids
(c)   Electro-chemical techniques

While general guidance is available [7.7, 7.10, 7.18, 7.32], emphasis here is put on EN 1504 (Table 7.12) and its Principles (Table 7.4).

7.11.3.1　Coatings and impregnations (category 3, Table 7.5)

The efficacy of surface treatments is dependent on the quality and porosity of the concrete – e.g. effective impregnation can be difficult with high-strength dense concretes. Surface condition is also important and many coatings require careful pre-preparation before application. Surface treatments are most effective if applied during the initiation phase of a deterioration mechanism and can have different performance requirements for different aggressive actions. Consideration should also be given to the nature and location of the structure, and to the environmental conditions at its surface; this applies in particular to the moisture state and the likely ambient temperature range – the former can affect adhesion and the latter, the necessary elasticity in the coating.

While the numerical procedures in Section 7.9 can again be used for preliminary selection, the range of products, and their different actions, is so large that greater precision and detail are required. This can best come out of the EN 1504 stable, where surface treatments are covered in EN 1504-2, but in concert with Part 9 (design) and Part 10 (execution). Another strength is Part 8 on quality control and evaluation of conformity, supported by numerous Standards on test methods (see Appendices 1 and 2).

The different types of surface treatment are given in Table 7.15, together with some indication of their basic action and a listing of the main base

Table 7.15 Types of surface treatment

Type of treatment	Basic action	Principal base materials	Relevant EN 1504 principles (Table 7.4)
Hydrophobic impregnation	Seals pores. Creates water-repellent surface	Silane or siloxane	P1, P2, P8
Impregnation	Blocks pores Strengthens concrete surface	Mainly acrylics or epoxies Can be inorganic (fluoride compounds)	P1, P5
Coating	Continuous layer as a barrier (0.1–5 mm thick)	Epoxy, acryl polyurethane	P1, P2, P5, P6, P8

materials. The last column is included to indicate the main use of each type; this is done by linking with the EN 1504 Principles (Table 7.4).

In making a selection, it is also important to take account of the type of defect involved and the dominant aggressive action, since this will indicate the primary EN 1504 Principle to be followed. Some guidance on that is given in Table 7.16.

The field of surface treatments is dynamic, with new or improved products continuing to appear on the market. In general, the action of these is fairly well understood, and the advent of EN 1504 has introduced a systematic approach to their performance requirements and evaluation. Their long-term efficacy in the field is less clearly established (see Section 7.7, and references [7.15–7.17]); the quality of site practice, including surface condition and preparation, is crucial for their effectiveness. Nor do short-term laboratory tests always give a true picture of their real deterioration with time, under the wide range of environmental conditions that can occur in practice. Surface treatments can also be vulnerable to physical effects such as abrasion or the influence of movement and strain.

It is important, therefore, at the selection stage, to establish links with repair contractors specialising in different techniques, who may be able to supply details of successful case studies – even give guarantees – and certainly give details of their operational and quality-control procedures.

Table 7.16 Examples of the main EN 1504 Principles involved with different aggressive actions

Defects and degradation processes	Principle	
	Related to defects in concrete	Related to reinforcement corrosion
Ingress of aggressive substances, e.g. chlorides, carbon dioxide and acids.	P1, P6	P8
Carbonation	P1	
ASR	P2	
Frost	P2	
Reinforcement corrosion		P8
Weak concrete surface	P5	

7.11.3.2 Filling cracks and voids (category 4, Table 7.5)

Some performance requirements for this type of repair are given in Sections 7.4.3 and 7.4.4. In EN 1504, this is covered in EN 1504-5 (Table 7.12), but with the emphasis on crack injection methods; i.e. the filling of small external surface voids is regarded as mini-patching in EN 1504, and Section 7.11.2.3 applies.

The primary purpose of crack injection is protection against the ingress of aggressive agents – Principle 1 in Table 7.4. It will provide some measure of moisture control (Principle 2), particularly if used with a coating. Table 7.4 perceives crack injection as contributing to structural strengthening, but this effect is generally minor, in restoring section integrity.

Injection materials can be based on either hydraulic or polymer binders. As for patch materials or coatings, there is a wide range of products on the market. While the Section 7.9 approach can again be used for preliminary selection, detailed evaluation can only come from an approach similar to that provided by EN 1504-5. Adhesion, shrinkage and workability properties are particularly important, since the cracks are generally relatively narrow and can penetrate beyond the location of reinforcement. As for all preventative measures, execution and workmanship standards are paramount, and the advice of specialist contractors essential at the selection stage and in the preparation of specifications for the work.

7.11.4 Level (4) in Figure 7.7 – Electro-chemical techniques

7.11.4.1 Introduction

Categories 1–4 in Table 7.5 represent the traditional approach to rehabilitation, i.e. directly addressing the effects of the aggressive actions via strengthening or protection. The electro-chemical approach is different, since it tackles the corrosion process itself, in terms of slowing down, interfering with it or, ideally, stopping it all together. Current practice recognises four ways of doing that, namely:

- re-alkalisation
- chloride extraction
- cathodic protection
- corrosion inhibitors.

To fully understand how each technique functions, it is necessary to have a working knowledge of the chemical processes that cause corrosion; references [4.10–4.15] provide that. Table 7.12 shows no particular EN 1504 Part on this approach – surprising perhaps in view of the emphasis given to corrosion in Principles 7–11 (Table 7.4) – but rather two separate CEN

Standards [7.33, 7.34]. Unlike other Standards for repair, these are very much process-related, rather than being product-based. General guidance is available, either on individual techniques [7.9] or on the overall approach [7.35]. Additionally, some feedback is available on field experience, e.g. reference [7.36].

The use of these methods may well increase in the future, as greater emphasis is put on the hands-on management and monitoring of structures. It is important to have an overall perspective of their role. To achieve that, reference is made to Figure 7.6 on the timing of an intervention. With electro-chemical techniques, the earlier the intervention, the greater the benefit obtained (zones 1 and 2 in Figure 7.6). To emphasis that point, reference is also made to the two-phase corrosion model due to Tuutti [4.17], reproduced here as Figure 7.9.

Superimposed on the two-phase initiation and propagation line are two possible intervention points, marked proactive and reactive. During the initiation phase, corrosion has not yet started, but either carbonation or chloride fronts are advancing towards the reinforcement. Re-alkalisation or chloride extraction treatments, or possibly corrosion inhibitors, suggest themselves, should the owner wish to avoid or delay the onset of corrosion.

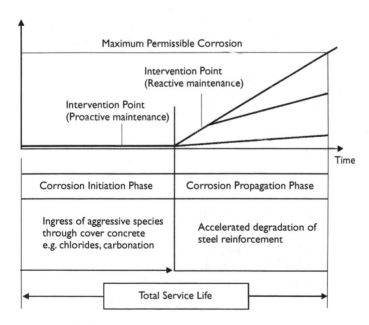

Figure 7.9 Simplified two-phase corrosion model (Tuutti [4.17], reproduced with permission from [7.16]. IHS BRE Press, 2007)

If intervention is to take place during the propagation phase, some corrosion will already have occurred, and the objective is ideally to stop further corrosion or at least to slow down the corrosion rate (one of the flatter gradient lines in the diagram). Electro-chemical techniques can do that, while also coping with the risk of new corrosion cells starting up, as can occur on the fringe of repaired areas when physical restoration methods are used.

All electro-chemical methods have principles and practical details in common, the main differences being the magnitude of the current and the duration of the treatment to meet the different objectives. A simplified representation of the different processes that occur is shown in Figure 7.10.

An external anode is provided and a current flows through the concrete, via an electrolyte, to the reinforcement, which acts as the cathode in an electro-chemical cell. By appropriate design of the system, the effect can be:

- reinstatement of the alkalinity of the pore solution – re-alkalisation (RA in the diagram);
- removal of the aggressive chloride ions – chloride extraction (CE in the diagram);
- re-passivation of the reinforcement to a more negative potential – cathodic protection (CP in the diagram).

Figure 7.10 is a simplification of the processes involved in each method. Some additional information is given in Sections 7.11.4.2–7.11.4.4 on basic principles, while recognising that there are variations possible within each, whose details can be found in the references given to this section.

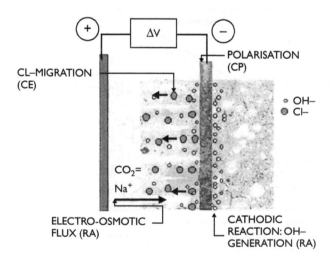

Figure 7.10 Schematic diagram of the different processes that occur with electro-chemical techniques

7.11.4.2   Re-alkalisation and chloride extraction

The methods used for both processes are similar and can be described together. Pre-preparation is important, in terms of cleaning the reinforcement, and removing cracked concrete and replacing it with material of similar electrical resistivity.

Both are temporary treatments, re-alkalisation lasting 1–2 weeks, and chloride extraction, anything between 3 and 15 weeks, depending on the chloride profile and concentration. Process parameters are current density and treatment time. An impressed current is applied, via a direct-current electricity supply whose positive terminal is connected to the temporary anode and the negative terminal, to the cathode (reinforcement). Electrolytes vary, generally being solutions of sodium or potassium carbonate for re-alkalisation and of calcium hydroxide for chloride extraction.

For re-alkalisation, the formation of hydroxyl ions via the electric current raises the alkalinity at the level of the reinforcement, with a pH value of 10–11 being the minimum target. For chloride extraction, negatively charged chloride ions are attracted to the external anode and repelled by the cathode, and either end up in the electrolyte or get positioned away from the reinforcement.

In practice, there are variations on the market in terms and electrolyte, and, in making judgements, it is the efficacy of the process that has to be assessed, in consultation with experienced contractors, and, hopefully, with evidence from case studies and the laboratory. The effects appear to be quite long-lasting, and again evidence is required of the likely duration.

There are potential hazards which also need to be considered. These include the following:

- The increased alkalinity may lead to ASR, if reactive aggregates are present.
- Both processes produce hydrogen at the reinforcement, which, for certain types of steel, may lead to hydrogen embrittlement.
- The processes are not normally recommended for high tensile steel such as that used for prestressing.
- The pore structure may be affected adjacent to the reinforcement, but is unlikely to cause bond problems with deformed bars.

7.11.4.3   Cathodic protection

This is a permanent treatment, involving either an impressed current or a sacrificial anode system; with the former, the anode is generally of titanium, with the latter, either zinc or aluminium. Impressed-current system is generally more expensive, also requiring monitoring and maintenance, but is more reliable in the longer term; keeping the chosen current constant is important. Sacrificial anode systems are easier to apply, but monitoring is more difficult.

The level of protection provided depends on chloride content, pH level and cement content and type. The permanency of the approach makes it attractive, when re-contamination is a real possibility. The technique is process-related, and this is reflected in the BS EN [7.33], in setting performance requirements.

7.11.4.4 Corrosion inhibitors

The use of corrosion inhibitors in the concrete mix for new construction has been the most common application, but retro-application as a curative measure is also possible. This is relatively new in the UK, compared with America, reflected in the relative vintage of available general guidance [7.37, 7.38].

There are several generic types on the market, with different modes of action. They come in the form of liquids, gels or powders, introduced into the concrete either on the surface or in discretely drilled holes, to reduce or interfere with the corrosion process. The most common types are either anodic (controlling the anodic reaction in the corrosion cell) or ambiotic, which address both the anode and the cathode. They are made up of complex chemical blends, probably generating more than one inhibiting mechanism. Dosage is crucial, as is its maintenance at the prescribed level over a period of time.

The use of inhibitors is not yet within the scope of EN 1504, except possibly for Part 10, dealing with site application and quality control. As experience and feedback is obtained, more detailed guidance may be expected, e.g. on The Concrete Repair Association website [7.10].

7.12 Concluding remarks

This chapter does not attempt a detailed treatment of all the options available for protection, prevention, repair or strengthening. It does try to provide an overall perspective and sources where more information can be found. The scene is constantly changing and will continue to do so.

What is being observed is a move away from empirical selection methods towards a more scientific basis, i.e. from Section 7.9 towards Section 7.10. Owners are becoming more demanding with higher expectations, and this trend will help meet their needs.

Section 7.7 is important in emphasising the need for more feedback on repairs in service, under the wide range of practical and environmental conditions that can occur. The timing of an intervention is also important; prevention is always better than cure – and generally cheaper in whole-life costing terms.

APPENDIX 7.1 CEN Standards in support of EN 1504 for protection and repair materials

European Standard test methods for protection and repair materials		Status	Coating & surface treatment	Repair mortars	Structural bonding	Injection products	Anchoring products	Reinforcement protection
Standard¹	Title		1504-2	1504-3	1504-4	1504-5	1504-6	1504-7
BS EN 1542	Measurement of bond strength by pull-off	P						
BS EN 1543	Determination of tensile strength development for polymers	P				√		
1544	Determination of creep under tensile stress at 23°C and 50°C	due					√	
BS EN 1766	Reference concretes for testing	P	√	√	√	√		
BS EN 1767	Infrared analysis	P	√					
BS EN 1770	Determination of coefficient of thermal expansion	P			√			
1771	Determination of injectability: wet medium – dry medium	P				√	√	
BS EN 1799	Tests to measure the suitability of structural bonding agents for application to concrete surface	P			√			
BS EN 1877-1	Reactive functions related to epoxy resins – Part 1: Determination of epoxy equivalent	P	√					
BS EN 1877-2	Reactive functions related to epoxy resins – Part 2: Determination of amine functions using the total basicity number	P	√					
1881-1	Pull-out test – Part 1: Uncracked concrete	P					√	
BS EN 12188	Determination of adhesion steel-to-steel for characterisation of structural bonding agents	P			√			
BS EN 12189	Determination of open time	P			√			
BS EN 12190	Determination of compressive strength of repair mortar			√				

European Standard test methods for protection and repair materials

Standard[1]	Title	Status	Coating & surface treatment 1504-2	Repair mortars 1504-3	Structural bonding 1504-4	Injection products 1504-5	Anchoring products 1504-6	Reinforcement protection 1504-7
BS EN 12192-1	Granulometry analysis – Part 1: Test method for dry components of premixed mortar	P		✓				
BS EN 12192-2	Granulometry analysis – Part 2: Test method for fillers for polymer bonding agents	P			✓			
BS EN 12614	Determination of glass transition temperature of polymers	P						✓
BS EN 12615	Determination of slant shear strength	P			✓			
BS EN 12617-1	Part 1: Determination of linear shrinkage for polymers and surface protection systems	P	✓					
BS EN 12617-2	Shrinkage of crack injection product formulated with polymer binder – Part 2: Volumetric shrinkage	P						✓
BS EN 12617-3	Part 3: Determination of early age linear shrinkage for structural bonding agents	P			✓			
BS EN 12617-4	Part 4: Determination of shrinkage and expansion	P		✓				
BS EN 12618-1	Adhesion and elongation capacity of injection products, with limited ductility	P						✓
BS EN 12618-2	Determination of the adhesion of injection products, with or without thermal cycling – Part 2: Tensile bond method	P						✓
BS EN 12618-3	Determination of the adhesion of injection products, with or without thermal cycling – Part 3: Slant shear method	P						✓
BS EN 12636	Determination of adhesion concrete to concrete	P			✓			
BS EN 12637-1	Compatibility of injection products – Part 1: Compatibility with concrete	P						✓

Reference	Description	P			
BS EN 12637-3	Compatibility of injection products – Part 3: Effect of injection products on elastomers	P	✓		
BS EN 13057	Determination of resistance of capillary absorption	P		✓	
BS EN 13062	Thixotropy	P			✓
BS EN 13294	Determination of stiffening time	P		✓	
BS EN 13295	Determination of resistance to carbonation	P		✓	
BS EN 13395-1	Determination of workability – Part 1: Test for thixotropic repair mortars	P		✓	
BS EN 13395-2	Determination of workability – Part 2: Test for flow of grout or mortar	P		✓	
BS EN 13395-3	Determination of workability – Part 3: Test for flow of repair concrete	P		✓	
BS EN 13395-4	Determination of workability – Part 4: Application of repair mortar overhead	P		✓	
BS EN 13396	Measurement of chloride ion ingress	P		✓	
BS EN 13412	Determination of modulus of elasticity in compression	P		✓	
BS EN 13529	Resistance to severe chemical attack	P			✓
BS EN 13578	Compatibility on wet concrete	P			✓
BS EN 13579	Drying test for hydrophobic impregnation	P			✓
BS EN 13580	Water absorption and resistance to alkali for hydrophobic impregnations	P			✓
BS EN 13581	Determination of loss of mass of hydrophobic impregnated concrete after freeze–thaw salt stress	P			✓
BS EN 13584	Creep in compression	P		✓	
BS EN 13687-1	Determination of thermal compatibility – Part 1: Freeze–thaw cycling with de-icing salt immersion	P		✓	
BS EN 13687-2	Determination of thermal compatibility – Part 2: Thunder shower cycling (thermal shock)	P		✓	
BS EN 13687-3	Determination of thermal compatibility – Part 3: Thermal cycling without de-icing salt impact	P		✓	

APPENDIX 7.1 (Continued)

European Standard test methods for protection and repair materials

Standard[1]	Title	Status	Coating & surface treatment 1504-2	Repair mortars 1504-3	Structural bonding 1504-4	Injection products 1504-5	Anchoring products 1504-6	Reinforcement protection 1504-7
BS EN 13687-4	Determination of thermal compatibility – Part 4: Dry thermal cycling	P		✓				
BS EN 13687-5	Determination of thermal compatibility – Part 5: Resistance to temperature shock	P	✓					
BS EN 13733	Determination of the durability of structural bonding agents	P			✓			
BS EN 13894-1	Determination of fatigue under dynamic loading – Part 1: During cure	P			✓			
BS EN 13894-2	Determination of fatigue under dynamic loading – Part 2: After hardening	P			✓			
BS EN14068	Determination of watertightness (of injection products)	P						✓
BS EN 14117	Determination of viscosity of cementitious injection products	P				✓		
BS EN 14406	Determination of the expansion ratio and expansion evolution	P				✓		
BS EN 14497	Determination of the filtration stability	due				✓		
BS EN 14498	Volume and weight changes after air drying and water storage cycles	due				✓		
BS EN 14629	Determination of chloride content in hardened concrete	due						✓
BS EN 14630	Determination of carbonation depth in hardened concrete by the phenolphthalein method	due						✓

Notes

1. This appendix is based on a text downloaded from the Concrete Repair Association website www.cra.org.uk. The text was prepared by the Hywel Davies Consultancy, with an effective date of 2 October 2006.
2. The relevance of each Standard is shown in relation to the appropriate Part of EN 1504. 'P' denotes 'published'.

APPENDIX 7.2 A selection of ISO Standards of relevance to different Parts of EN 1504

Standard (all prEN unless stated otherwise)	Title	Coating & surface treatment	Repair mortars	Structural bonding	Injection products	Anchoring products	Reinforcement protection
		1504-2	1504-3	1504-4	1504-5	1504-6	1504-7
EN ISO 1517	Paints and varnishes – Surface-drying test. Ballotini method	✓					
EN ISO 2409-6	Method for tests for paints – Part 6: Cross cut tests	✓					
EN ISO 2808	Paints and varnishes – Determination of film thickness (ISO 2808:1997)	✓					
ISO 2811-1	Methods of test for paints – Determination of density by the pyknometer method. Also available as BS 3900-A19:1998	✓					
ISO 2811-2	Methods of test for paints – Determination of density by the immersed body (plummet) method. Also available as BS 3900-A20:1998	✓					
EN ISO 2812-1	Paints and varnishes – Determination of resistance to liquids (chemical resistance)	✓					
EN ISO 2815	Paints and varnishes – Buchholz indentation test (ISO 2815:1973)	✓					
EN ISO 3219	Determination of viscosity using a rotational viscometer with defined sheer rate	✓					

APPENDIX 7.2 (Continued)

Standard (all prEN unless stated otherwise)	Title	Coating & surface treatment	Repair mortars	Structural bonding	Injection products	Anchoring products	Reinforcement protection
		1504-2	1504-3	1504-4	1504-5	1504-6	1504-7
EN ISO 3251	Paints and varnishes – Determination of non-volatile matter of paints varnishes and binders for paints and varnishes	✓					
EN ISO 3274	Geometrical product specifications (GPS) – Surface texture: Profile method – Nominal characteristics of contact (stylus) instruments (ISO 3274:1996)	✓					
EN ISO 3451-1	Plastics – Determination of ash – General methods	✓					
prEN ISO 4628-2	Paints and varnishes – Evaluation of degradation of coatings – Designation of quantity and size of defects, and of intensity of changes – Part 2: Assessment of the degree of blistering	✓					
prEN ISO 4628-3	Paints and varnishes – Evaluation of degradation of coatings – Designation of quantity and size of defects, and of intensity of changes – Part 4: Assessment of the degree of flaking	✓					
prEN ISO 4628-4	Paints and varnishes – Evaluation of degradation of coatings – Designation of quantity and size of defects, and of intensity of changes – Part 4: Assessment of degree of cracking	✓					

prEN ISO 4628-5	Paints and varnishes – Evaluation of degradation of coatings – Designation of quantity and size of defects, and of intensity of changes – Part 5: Assessment of the degree of flaking	✓
prEN 4628-6	Paints and varnishes – Evaluation of degradation of paint coatings – Designation of intensity, quantity and size of common types of defect – Part 6: Rating of degree of chalking by tape method	✓
EN ISO 5470-1	Rubber or plastic-coating fabrics – Determination of abrasion resistance. Part 1: Taber abrader	✓
BS EN ISO 6272	Paints and varnishes – Falling-weight test	✓
EN ISO 7783-1	Paints and varnishes – Determination of water vapour transmission rate – Part 1: Dish method for free films	✓
EN ISO 9514	Paints and varnishes. Determination of the pot-life of liquid systems. Preparation and conditioning of samples and guidelines for testing	✓

Note
1. This appendix is derived from a downloaded text from the Concrete Repair Association website www.cra.org.uk. This text was prepared by the Hywel Davies Consultancy, with an effective date of 2 October, 2006.

References

7.1 Perkins P.H. *Repair, protection and waterproofing of concrete structures*. E.H. Spon, UK. 3rd Edition. 1997.

7.2 Mallett G.P. *Repair of concrete bridges: state of the art review*. Thomas Telford Ltd, UK. 1994.

7.3 Highways Agency et al. *Post-tensioned concrete bridges. Anglo-French liaison report*. Thomas Telford Ltd, London. UK. 1999. p. 164.

7.4 The Concrete Society. *Strengthening concrete structures with fibre composites – acceptance, inspection and monitoring*. Technical Report 57. 2002. The Concrete Society, Camberley, UK.

7.5 The Concrete Society. *Construction and repair with wet-process sprayed concrete*. Technical Report 56. 2002. The Concrete Society, Camberley, UK.

7.6 The Concrete Society. *Design guidance for strengthening concrete structures using fibre composite materials*. Technical Report 55 2nd Edition. 2004. The Concrete Society, Camberley, UK.

7.7 The Concrete Society. *Guide to surface treatments for protection and enhancing durability*. Technical Report 50. The Concrete Society, Camberley, UK.

7.8 The Concrete Society. *Patch repairs of reinforced concrete subject to reinforcement corrosion*. Technical Report 38. 1991. The Concrete Society, Camberley, UK.

7.9 The Concrete Society. *Cathodic protection of reinforced concrete*. Technical Report 36. 1989. The Concrete Society, Camberley, UK.

7.10 The Concrete Society. *Enhancing reinforced concrete durability*. Technical Report 61. 2006. The Concrete Society, Camberley, UK.

7.11 Tilly G.P. Performance of repairs to concrete bridges. *Proceedings of the Institution of Civil Engineers. Bridge Engineering*. Vol. 157. pp. 171–174. September, 2004. ICE, London, UK.

7.12 REHABCON. EC Innovation and SME project EC DE ENTR-C-2. Strategy for maintenance and rehabilitation concrete structures. Deliverable D2 of Work Package 2.2. Available from the Building Research Establishment (BRE), Garston, UK, November, 2002.

7.13 CEN. Products and systems for the protection and repair of concrete structures – definitions, requirements, quality control and evaluation of conformity. EN 1504, Parts 1–10. CEN, Brussels.

7.14 Alexander M. et al. Concrete repair, rehabilitation and retrofitting. Taylor & Francis, London, UK. 2006. p. 511.

7.15 Tilly G.P. and Jacobs J. Concrete repairs: observations on performances in service and current practice. CONREPNET Draft Document for comment. Available from Building Research Establishment (BRE) Garston, UK. 2006.

7.16 CONREPNET. Achieving durable concrete structures; adopting a performance-based intervention strategy. First draft document, October 2006. pp. 225. Available from the project co-ordinator: Building Research Establishment (BRE), Garston, UK.

7.17 Baldwin N.J.R and Kine E.S. Field studies of the effectiveness of concrete repairs. Phase 4 report. Health & Safety Executive (HSE). UK. RR 186. HSE Books, UK. 2003.

7.18 Emmons P. and Matthews S.L. (editors). *Concrete repair manual*, 2nd Edition. International Concrete Repair Institute, USA. 2003.

7.19 LIFECON. Life cycle management of concrete infrastructures for improved sustainability. EC Competitive and Sustainable Growth Programme. Contract GIRD – CT – 2000 – 037B. Deliverables are located at website www.wtt.fi/rte/strat/projects/lifecon/summary.htm.

7.20 NORECON. Network on repair and maintenance of concrete structures. 3 technical reviews on: Decisions and requirements for repair; Repair methods; Guidelines for manufacturers, contractors and consultants on the basis of European Standards. Documents are located at website www.nordicinnovation.net.

7.21 REHABCON Manual – Strategy for maintenance and rehabilitation in concrete structures. Innovation and SME Programme – Contact IPS – 2000 – 0063. Website: www.cbi.se/rehabcon/rehabconfiles.

7.22 British Standards Institution (BSI). Eurocode 2: Design of concrete structures – Part 1.1: General rules and rules for buildings. BS EN 1992-1-1: 2004. BSI, London, UK. 2004.

7.23 Andrade C. and Izquierdo D. *Evaluation of best repair options through the repair index method RIM*. Paper contained in reference 7.14. Taylor and Francis, London, UK. 2006. pp. 283–284.

7.24 Highways Agency. *Steel plate bonding*. BA30/94. The Highways Agency, London. UK. 1994.

7.25 Allen R.T.L., Edwards S.C. and Shaw J.D.N. *The repair of concrete structures*, 2nd Edition. Blackie Academic & Professional, London, UK. 1994.

7.26 Mays G.C. and Hutchinson A.R. *Adhesives in civil engineering*. Cambridge University Press, UK. 1992.

7.27 Highways Agency. *Design of bridges and concrete structures with external and unbonded tendons*. BD and BA58/02. Highways Agency, London, UK. 1994.

7.28 Concrete Society. *Durable bonded post-tensioned concrete bridges*. Technical Report 47, 2nd Edition. 2000. Concrete Society, Camberley, UK.

7.29 Highways Agency. Design Manual for Roads and Bridges (DMRB). Website – http://www.official-documents.co.uk/documents/deps/ha/dmrb/index.htm.

7.30 Concrete Repair Association (GRA) www.cra.org.uk.

7.31 Sivfwerbrand J. Shear bond strength in repair concrete structures. *Materials and Structures*. Vol. 36, July, 2003. pp. 419–424.

7.32 Thomas H. *Handbook of coatings for concrete*. Whittles Publishing, Scotland, UK. 2002.

7.33 British Standards Institution (BSI). *Cathodic protection of steel in concrete*. BS EN 12696: 2000. BSI, London, UK. 2000.

7.34 CEN. *Electro-chemical re-alkalisation and chloride extraction treatments for reinforced concrete*. pr EN 14038. In draft form. CEN, Brussels.

7.35 Mietz J. Electro-chemical rehabilitation methods for reinforced concrete structures. European Federation of Corrosion (EFC) Publication 24. The Institute of Materials, London, UK. 1998.

7.36 Elsender J. et al. Repair of reinforced concrete structures by electro-chemical techniques – field experience. EFC Publication 25. The Institute of Materials, London, UK. 1998.

7.37 National Research Council (NRC). *Concrete bridge protection and rehabilitation : corrosion inhibitors and polymers*. Report SHRP-S-666. 1993. NRC, Washington, USA.

7.38 Mortlidge J.R. and Sergi G. *Corrosion inhibitors for reinforced concrete*. BRE Information Paper. BRE, Garston, UK. 2003.

Chapter 8

Back to the future

Using a somewhat hazy crystal ball, this chapter attempts to project ahead on future changes, based on a backward look at practice over the past 50 years or so.

The current annual costs of repair and remedial work in the UK is variously quoted as greater than 50 per cent of the output of the construction industry or £1 billion. As the concrete infrastructure ages, more structures will enter the renovation arena, and, since many current treatments have been shown to have relatively short lives (Section 7.7), further rounds of restoration are likely to be required in the longer term. Continuous asset management is here to stay, on a scale equal to, if not greater than, that for new construction. Under these circumstances, what changes might reasonably be expected, and what drivers will influence these?

8.1 Drivers

Apart from a general desire in the industry as a whole to 'do better', there are two major ones.

8.1.1 Client/owner demand

Owners have always been fairly comfortable with asset management in terms of routine maintenance and basic upgrading aimed at improving function and operations and the raising of standards generally (the obsolescence issue). They are less comfortable with unexpected increased costs due to a lack of durability over the expected lives of their structures. As a result, they are becoming more demanding, both for new build and remedial work, generally. A shift in emphasis is taking place (although by no means universal) from lowest first costs towards whole-life costing.

8.1.2 Sustainability

The industry is yet to come fully to terms with the concept of sustainability, where only the relatively easy targets of energy, emissions and waste have been addressed in any serious way. Infrastructure is a major consumer of natural resources. Obtaining a better fit for structures to meet owners' performance requirements for significant in-service periods is a major issue, while simultaneously minimising disruptions to operations caused by extensive repairs.

As outlined above, sustainability is a general issue, applying to new build and renovation alike. It is also likely that sustainability will come more into the selection process for the different repair options; some indication of how this might be done is given by Arya and Vassie [8.1]. The approach is similar in character to that given in Section 7.9 for repair options in general.

8.2 Design issues

There are several features here where change is certainly needed, if not yet occurring in any general way.

8.2.1 Assessment of 'loads'

Many aggressive actions have been identified and are fairly well understood. These include sulfate attack, ASR, abrasion and leaching. Material specifications exist to minimise the risk of these occurring.

However, feedback from performance in service has clearly shown the importance of considering the environment in design (Chapters 2 and 4 generally; Section 7.7 with respect to repair options). Traditional design methods require structural engineers to consider the effects of wind and temperature in their designs, yet the role of water is sadly neglected, except for calling up concrete grades [2.29]. Figures 2.5–2.9 show some of the micro-climate conditions that can occur in practice for different types of structure, and Table 2.2 indicates the importance of moisture in durability terms. It is suggested – strongly – that weight be given to water in design, comparable to that for temperature and wind.

8.2.2 Architectural and structural detailing

Following on from Section 8.2.1, a better balance has to be found between the current obsession with detailed structural analysis and need for structural detailing. This applies especially to all joints and those parts of the structure directly exposed to the local environment as Figure 2.2 clearly shows. In doing so, it is necessary to consider the mechanisms whereby moisture might penetrate into sensitive parts of the structure, e.g. Figures 2.4 and 2.9.

8.2.3   A holistic approach

The author feels strongly about this (reference [2.1]). Design and detailing must be linked to material specifications and to execution issues. Section 2.3 shows the importance of the latter, in recognising the variability that can occur on site with respect to concrete quality and cover. On cover, references [2.19–2.26] indicate an almost chronic inability to achieve specified levels; it is only relatively recently that designers have been encouraged to specify not only cover but also the methods of achieving it [8.2].

Much of this is due to poor workmanship and lack of quality-control on site, requiring method statements and training of all personnel. However, it is not solely due to that, and proper consideration of buildability issues is required at the design stage. In the development of repair systems via EN 1504, it is encouraging to note the emphasis put on a holistic approach, driven by performance requirements; the same approach is essential for new construction. There is little doubt that modern concretes are more durable, and developments in concrete technology are also beneficial, e.g. self-compacting concrete, but it is the totality of the overall package that matters at the end of the day.

8.3   Evolution of repair systems – the EN 1504 influence

While all ten parts of EN 1504 are now published (Table 7.12), its use in practice is still at a relatively early stage, with more supporting standards still to be put in place (Appendices 7.1 and 7.2). It will become more significant in the future, and there is little doubt that the repair industry is responding positively; reference [8.3] gives a particular example. Going beyond that, there is also equipment development for testing repair materials and systems, in support of the EN 1504 principles; reference [8.4] gives a particular example.

This is highly encouraging, provided all Parts of EN 1504 are applied, together with its Principles, i.e. a total package approach – design, materials, execution, quality assurance and acceptance testing.

8.4   Testing techniques

Test methods have improved beyond all recognition in recent years, particularly non-destructive testing (NDT) techniques (Table 8.1). A flavour of this is given in Chapter 4, with a listing appearing in Table 4.10, in support of the preliminary diagnosis issues in Table 4.9. Up-to-date information on this is contained in reference [8.5], where it is illuminating to compare this fourth edition with the first edition dating back decades.

Table 8.1 Available non-destructive test methods – Grantham [8.6]

Property under investigation	Test	Equipment Type
Corrosion of embedded steel	Half-cell potential	Electrochemical
	Resistivity	Electrical
	Linear polarisation resistance	Electrical
	AC Impedance	Electrochemical
	Cover depth	Electromagnetic
	Carbonation depth	Chemical/microscopic
	Chloride concentration	Chemical/electrical
	Surface hardness	Mechanical
Concrete quality, durability and deterioration	Ultrasonic pulse velocity	Electromechanical
	Radiography	Radioactive
	Radiometry	Radioactive
	Neutron absorption	Radioactive
	Relative humidity	Chemical/electronic
	Permeability	Hydraulic
	Absorption	Hydraulic
	Petrographic	Microscopic
	Sulfate content	Chemical
	Expansion	Mechanical
	Air content	Microscopic
	Cement type and content	Chemical/microscopic
	Abrasion resistance	Mechanical
Concrete strength	Cores	Mechanical
	Pull-out	Mechanical
	Pull-off	Mechanical
	Break-off	Mechanical
	Internal fracture	Mechanical
	Penetration resistance	Mechanical
	Maturity	Chemical/electrical
	Temperature-match curing	Electrical/electronic
Integrity and performance	Tapping	Mechanical
	Pulse-echo	Mechanical/electronic
	Dynamic response	Mechanical/electronic
	Acoustic emission	Electronic
	Thermoluminescence	Chemical
	Thermography	Infrared
	Radar	Electromagnetic
	Reinforcement location	Electromagnetic
	Strain or crack measurement	Optical/mechanical/electrical
	Load test	Mechanical/electronic/electrical

The impact that this could and should have on attitudes and practices is stressed by Grantham [8.6]. Quite apart from major improvements in traditional NDT techniques whose application is covered in Table 4.11, new techniques can provide much more detailed and relevant information, e.g. it is now possible to detect whether reinforcement is still corroding or not, and hence whether a repair is working successfully.

The primary role of NDT methods has been investigative, in support of a preliminary diagnosis. These new techniques suggest that this role could be expanded positively into a continuous monitoring regime. This possibility has recently been investigated in some depth by CIRIA [8.7]. Automated data handling methods make this a real possibility for wide use in the future, giving much more control to the whole asset management process.

8.5 Developments in asset management systems

Chapter 3 gave a brief review of developments in this area, while recognising four sources of guidance:

1. National and international guidance
2. Structure-specific approach
3. Recommendations from work on specific aggressive actions
4. Guidance from work on inspection and testing procedures

Items 3 and 4 provide the raw data for assessment purposes, and the outputs are fed into items 1 and 2. Item 1 is concerned with shaping both the framework and contents of asset management systems, which may be formal (e.g. British & ISO Standards) or informal (outputs from 'voluntary' bodies such as The Concrete Society, *fib*, or ACI).

It is in item 2, that the development of asset management systems per se has taken place. Here, it is reasonable to expect that different systems are necessary for different types of structure, e.g. for car parks, compared with nuclear installations. It is in the field of bridges that most work has been done (through sheer necessity), and, in the UK, it is noted that the numerous Standards and Advice Notes issued by the Highways Agency have been given focus by a Code of Practice [3.22].

The difficulty with structure-specific approaches is keeping them up to date, as more information is gained or new general guidance documents emerge. Major owners of structure populations have now moved towards computer-based systems capable of interrogation. As part of that, it is essential to have modules, which might be classified as 'knowledge databases', where new information is evaluated, and, if appropriate, fed into the action or processing modules. The SAFEBRO system has such a module (Figure 3.5) and future developments should see this approach applied more widely.

The above is a hoped-for development, systematic in nature. A further desirable development is technical. Throughout Chapters 5 and 6, the need to be clear on what constitutes minimum acceptable technical performance has been stressed. In Chapter 6, the approach is to calculate this by using design equations modified to allow for the effects of deterioration, and taking account of measured values for loads and mechanical properties. This is simple to do, and permits direct comparison with what was provided in the original design by empirical or semi-probabilistic methods. It can also be justified as being compatible with the quality and quantity of the necessary input.

There is room for improvement. In research terms, the current emphasis is on reliability and probabilistic methods in assessing safety levels. In a sense, this is a parallel to what was done for original design (see Figure 6.2). However, it has to be done with some care in assessment, for a number of reasons:

- the quality and quantity of the necessary inputs;
- structural sensitivity; the possibility of hidden strengths and alternative load paths; real boundary conditions;
- the quality of the as-built construction as a whole, plus local variations in concrete strength or reinforcement detailing;
- what is known about the structure overall from design drawings and previous inspections.

The Canadian Bridge Code [3.11] has attempted to address some of these issues with its back-to-basics approach involving reliability indices, as may be seen in Table 3.2. In the author's opinion, this has real potential, and requires further development, involving calibration against both deterministic and fully probabilistic methods (see Figure 6.2).

8.6 Concluding remarks

Because of the real need, it is reasonable to expect developments in all section of the assessment and management process. In previous sections in this chapter, some indication is given of where these might occur – or where they are needed.

Above all, there has to be balance in the overall package. Much is now known about deterioration mechanisms, but less about the effects that they produce under real environmental conditions in service. This suggests a greater emphasis on on-site measurement, in research terms. The assessment of residual structural capacity also needs above-average attention, as Chapter 6 indicates. The repair world is developing apace, and following the EN 1504 principles and approach should ensure that it moves in the right direction.

One might also express the almost pious hope of more technology transfer, with what is being learned from assessment work generally being transported back to improve the whole process for creating new construction. We have to stop making the same mistakes. With new build, current practice is not fully holistic, with improvements needed most in design, detailing and execution.

References

8.1 Arya C. and Vassie P.R. Assessing the sustainability of methods of repairing concrete bridges subjected to reinforcement corrosion. *International Journal of Materials and Product Technology*, Vol. 23, No. 3/4 Interscience Enterprises Ltd. 2005.

8.2 Shaw C. Cover to reinforcement – getting it right. *The Structural Engineer*, Vol. 85, No. 4. 20 February 2007. pp. 31–35. ISructE, London, UK.

8.3 Threadgold M. and Williams B. *Concrete repair strategies – The provisional European Standard and advanced repair mortar development concrete*, Vol. 40, No. 4. May 2006. pp. 30–31. The Concrete Society, Camberley, UK.

8.4 Meldrum V. Field testing equipment for coatings applied to concrete. *Concrete*, Vol. 40, No. 3. April 2005. pp. 36–37. The Concrete Society, Camberley, UK.

8.5 Bungey J., Millard S. and Grantham M. *Testing of concrete in structures*. 4th Edition. Taylor and Francis, London, UK. 2006.

8.6 Grantham M. Developments in non-destructive testing in the concrete industry. *Concrete*, Vol. 40, No. 3. April 2006. pp. 32–33. The Concrete Society, Camberley, UK.

8.7 CIRIA. *Intelligent monitoring of concrete structures*. Publication C6661. CIRIA, London, UK. 2007.

Index

Milton Keynes UK
Ingram Content Group UK Ltd.
UKHW021627071024
449327UK00020BA/1225